FALKLANDS ARMOURY

Blandford Press
Poole · Dorset

Edited by: Mark Dartford

First published in the U.K. 1985 by Blandford Press,
Link House, West Street, Poole, Dorset BH15 1LL

Copyright © Marshall Cavendish Ltd

Distributed in the United States by
Sterling Publishing Co., Inc.,
2 Park Avenue, New York, N.Y. 10016

Falklands armoury.
  1. Falkland Islands War, 1982   2. Great Britain
  —Armed Forces—Equipment—History
  I. Dartford, Mark
  623'.0997'11   UC46.G7

ISBN 0-7137-1691-6

Printed in Spain by Jerez Industrial S.A.

Contents

Introduction

When a small group of unconvincing scrap metal merchants jauntily hoisted a foreign flag over a tiny British dependency in the remote waters of the South Atlantic in 1982, few could have predicted any of the events this cheeky gesture of defiance would trigger off.

Within one year, a powerful South American military dictatorship had been replaced by a democratically-elected Government; two British Foreign Secretaries had been replaced, and the most unpopular Government since the war was surging ahead in the polls towards a second term of office, won with an outstanding majority. A French munitions company suddenly found itself inundated with orders for a hitherto little-known missile; a heavily-criticised British jump-jet suddenly became a much sought after machine, and the United States found itself acting as go-between on behalf of two belligerent nations it considered political allies. Even three years on, the controversies continue to rage, and political careers may yet hang on the balance of decisions and actions taken at the time.

Much has been written and broadcast about the events of those days: the heroes, the tragedies, the politics and the military strategy and tactics. *Falklands Armoury*, however, is not an attempt to find answers to, or explanations, for the conduct of the war. Rather, it is an illustrated tribute to the courageous men who travelled half way round the world and returned with what was – without question – a remarkable British military victory.

The Harrier: one reason why

The extraordinary virtues of the Harrier were one key to Britain's determination to contest the Argentine invasion. Those who knew the machine's potential were confident that it was an aircraft and a weapons system second to none.

'I THINK that some of the "experts" rather exaggerate the value, the strength of the Argentinian Air Force', commented Defence Secretary John Nott, early in the conflict. 'The Mirage aircraft really are a decade behind the Harrier, and beyond that their other aircraft are not particularly modern . . .

and the Harrier is a very effective aircraft . . .'

Harrier Chief Designer, John Fozzard, would certainly agree with Nott's criticism of 'experts'. Fozzard has his own definition of expert—' "Ex" is a "has been"—"spurt" is a drip under pressure!' But it was not just the age of Argentina's 44 Mirages (more than half of which were in fact Israeli-built Daggers), nor indeed the age of Argentina's three squadrons of Skyhawks that caused Nott's bold statements. It was to do with two vitally important differences—one between the different aircraft, and the other between two versions of the same missile carried by all three types of aircraft.

HARRIER

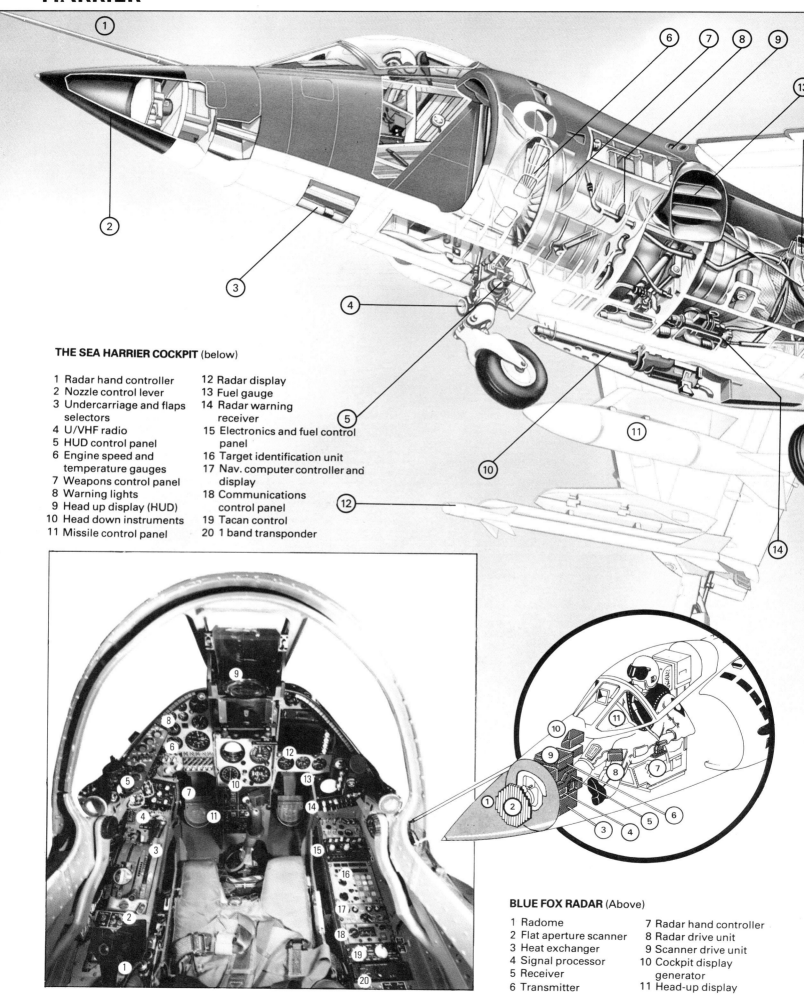

THE SEA HARRIER COCKPIT (below)

1 Radar hand controller
2 Nozzle control lever
3 Undercarriage and flaps selectors
4 U/VHF radio
5 HUD control panel
6 Engine speed and temperature gauges
7 Weapons control panel
8 Warning lights
9 Head up display (HUD)
10 Head down instruments
11 Missile control panel
12 Radar display
13 Fuel gauge
14 Radar warning receiver
15 Electronics and fuel control panel
16 Target identification unit
17 Nav. computer controller and display
18 Communications control panel
19 Tacan control
20 1 band transponder

BLUE FOX RADAR (Above)

1 Radome
2 Flat aperture scanner
3 Heat exchanger
4 Signal processor
5 Receiver
6 Transmitter
7 Radar hand controller
8 Radar drive unit
9 Scanner drive unit
10 Cockpit display generator
11 Head-up display

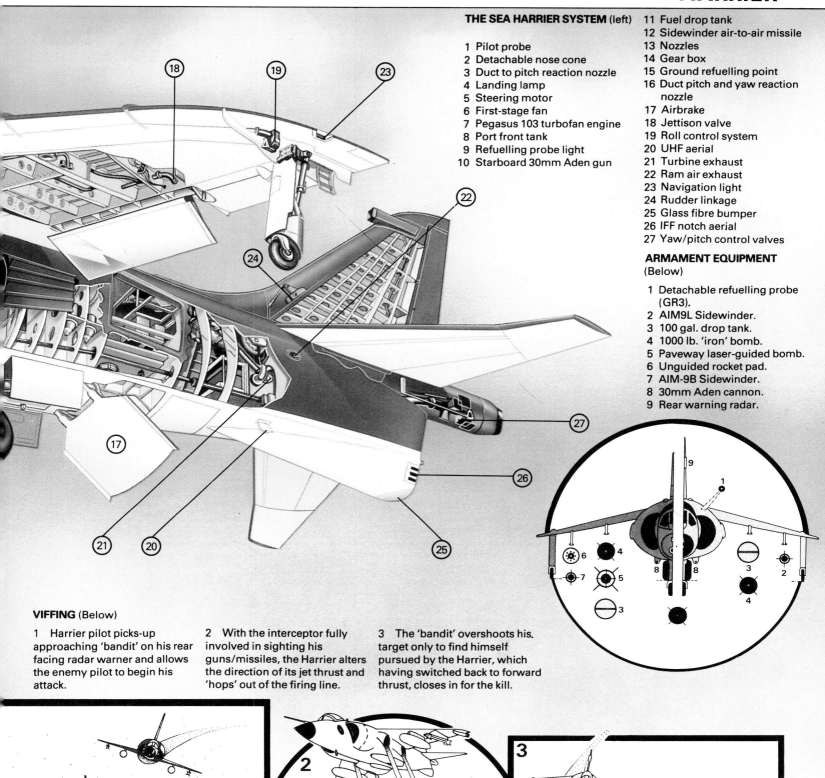

THE SEA HARRIER SYSTEM (left)

1 Pilot probe
2 Detachable nose cone
3 Duct to pitch reaction nozzle
4 Landing lamp
5 Steering motor
6 First-stage fan
7 Pegasus 103 turbofan engine
8 Port front tank
9 Refuelling probe light
10 Starboard 30mm Aden gun
11 Fuel drop tank
12 Sidewinder air-to-air missile
13 Nozzles
14 Gear box
15 Ground refuelling point
16 Duct pitch and yaw reaction nozzle
17 Airbrake
18 Jettison valve
19 Roll control system
20 UHF aerial
21 Turbine exhaust
22 Ram air exhaust
23 Navigation light
24 Rudder linkage
25 Glass fibre bumper
26 IFF notch aerial
27 Yaw/pitch control valves

ARMAMENT EQUIPMENT
(Below)

1 Detachable refuelling probe (GR3).
2 AIM9L Sidewinder.
3 100 gal. drop tank.
4 1000 lb. 'iron' bomb.
5 Paveway laser-guided bomb.
6 Unguided rocket pad.
7 AIM-9B Sidewinder.
8 30mm Aden cannon.
9 Rear warning radar.

VIFFING (Below)

1 Harrier pilot picks-up approaching 'bandit' on his rear facing radar warner and allows the enemy pilot to begin his attack.

2 With the interceptor fully involved in sighting his guns/missiles, the Harrier alters the direction of its jet thrust and 'hops' out of the firing line.

3 The 'bandit' overshoots his target only to find himself pursued by the Harrier, which having switched back to forward thrust, closes in for the kill.

HARRIER

Vectored thrust

The Harrier is the only British aircraft that could have gone to the South Atlantic with the fleet. And that is because of its unique ability to vector its thrust. HMS *Hermes* used to operate long range Buccaneer strike aircraft, and had it not been for a radical change in defence thinking, it would still be carrying the catapult and arrestor equipment used to launch and recover conventional fighters.

The addition of Buccaneers to the combat zone would have given Britain the capability to destroy the Argentine air bases at any time, day or night, during the conflict. However, without landing arrestor wires, the only way to recover an aircraft on a ship the size of *Hermes* or *Invincible* is by vertical landing, and only the Harrier is capable of vertical landing.

The Harrier's ability to change the direction of its engine thrust in flight was discovered by the US Marine Corps several years ago. The effect it has is to reduce the forward speed of the Harrier by hundreds of feet per second and at the same time increase the aircraft's altitude at an equally rapid rate of climb. Any conventional fighter trying to attack the Harrier, even it if shuts off all power, will still overshoot and in a matter of just a few seconds will come in range of the Harrier's weapon systems.

The combat situation between the Harriers and the Argentine Mirage, Dagger and Skyhawk fighters was made even more unfavourable for the Argentines because, unbeknown to them, Britain had secretly dipped into NATO stocks and equipped the Harriers with the latest version of the Sidewinder—the AIM9L. All of the Argentine aircraft were equipped with the earlier version of the missile which can only be fired from behind the target. That meant that the Harriers had to be attacked from behind on every occasion. The Harriers, however, could fire their AIM9L version from any aspects as this later model of Sidewinder will track a target from any angle.

If the Harrier is attacked, its own IFF (Interrogation Friend or Foe) system will tell the pilot from which direction he is being attacked. He then turns his aircraft in the same direction, moves the thrust vector level backwards and 'VIFF's the aircraft. (VIFF denotes 'vectoring in forward flight'). He could at this stage fire a Sidewinder AIM9L 'blind', as it will pick up the attacking aircraft as it flies under the now much higher Harrier.

Yet, before the Falklands War, the operational capabilities of the Royal Navy's Sea Harriers and the Royal Air Force's Harrier GR3s were being doubted by some, who regarded the aircraft as a star performer in air shows but not worth serious attention as an effective combat aircraft. The results of the air battles in the Falklands campaign have consigned such views to oblivion.

In the eyes of the world, the Harrier is now a fully fledged and proven combat aircraft. It deservedly ranks with other famous combat-proven names in UK aviation history, such as the Camel, Hurricane, Spitfire, Mosquito, Typhoon and Tempest—many of these also originating from the fighter design team which has been continuously in business at Kingston-upon-Thames in Surrey since 1913.

Heavily outnumbered by supersonic and subsonic aggressor aircraft, Harriers turned back or destroyed waves or Argentine Air Force A4 Skyhawks and Mirages during operations by the Task

Force. No Harrier was lost in air-to-air combat against the loss of nearly 30 attacking aircraft—the majority of them being Mirages and Skyhawks.

For the Harrier's creators, 'Operation Evita' (as it became unofficially known at Kingston) marked the turning point between belief and total conviction. Not only did the Harrier win its spurs, but the conflict also forced the urgent realization of several unusual Harrier-associated proposals which have been under discussion for some years.

One of these was the practicality of carrying—and perhaps operating—Harriers on merchant ships. This proposal is known under the name Arapaho.

The Arapaho concept aims to provide an aircraft carrier out of an existing suitable standard merchant ship—ideally a container ship which features a long flat upper deck with superstructure aft—by use of rapidly rigged decking bolted to the top layer of containers.

Because of its unobstructed length of deck, the container ship lends itself very well to conversion as an auxiliary VSTOL and helicopter carrier. Such a ship would, it had been argued, augment the fleet's carrier force in war at a fraction of the cost of a standard 'flat top', and could be made rapidly available at short notice.

This theory became fact when the Cunard container ship *Atlantic Conveyor* was chosen for the task. Taken out of her normal trans-Atlantic commercial service, *Atlantic Conveyor* sailed from her home port of Liverpool to HM Dockyard Devonport, in Plymouth, where in a matter of only a few days she was converted to carry Harriers and helicopters. Within her spacious hull, *Atlantic Conveyor* was loaded with vital supplies for the Task Force.

Atlantic Conveyor differed from the Arapaho concept in that she did not feature a full-length flight deck complete with ski-jump ramp, as would be the case in a complete conversion. Nevertheless, she did prove that the Arapaho concept was sound.

Tragically, *Atlantic Conveyor* was lost when hit by an air-launched Exocet missile delivered by an Argentine naval Super Etendard. Fortunately, all the Harriers she had been carrying on her special flight deck had been flown off and were being deployed from the Task Force carriers *Invincible* and *Hermes*, or were operating from sites on land. A large quantity of stores, however, was lost with the ship, including a number of heavy-lift RAF Chinook helicopters.

But all that was in the future. At the time, a topic even more absorbing, and much more pressing, than the potential of the Harrier was how to transport men and equipment to a hostile environment 8000 miles from home.

Extended ferry flights

Another important feature explored during the campaign was the Harrier's capability for extended ferry flights. Royal Navy and Royal Air Force VSTOL ferry flight records were broken when both Sea Harriers of 809 Squadron and Harriers of 1 Squadron flew from the UK to Ascension Island—a journey of some 4000 miles—refuelling in mid-flight from RAF Victor tankers.

Some Harriers subsequently flew on from Ascension Island direct to the fleet in latitude 52 degrees south—again a flight of about 4000 miles—vertically landing on the decks of the two carriers.

Below A Sea Harrier VSTOL takes off from the ski jump of HMS *Invincible*. Bottom An RAF Harrier GR3 firing 68mm SNEB rockets. Between them,

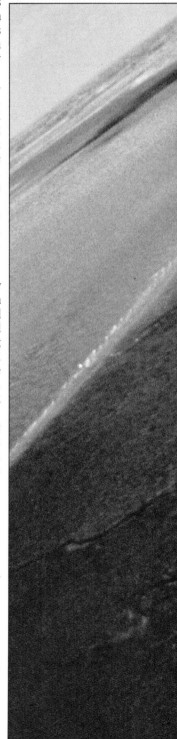

the two types accounted for 28 Argentine aircraft in aerial combat, without loss. Five Harriers fell to Argentine ground fire in all.

This was the first time that some of the pilots had landed on a ship at sea, which says volumes for the flexibility of VSTOL procedures in the Harrier.

Other Harriers arrived with *Atlantic Conveyor* where they reinforced units already engaged in air defence and ground attack roles.

A further development was the rapid conversion of RAF Harrier GR3s to carry Sidewinder air-to-air missiles, to enhance the Air Group's air defence capabilities while still retaining intact the ground attack force.

The Falklands crisis also accelerated the formation of the Royal Navy's third front-line operational Sea Harrier unit, 809 Squadron, which was formed at Royal Naval Air Station Yeovilton. This new unit was originally scheduled to operate from *Illustrious*, the RN's latest carrier and second ship in the *Invincible* class.

Not only was 809 Squadron formed rapidly—embracing a larger number of aircraft than its sister

front-line units—but also work on *Illustrious* was accelerated at the Walker Yard of Swan Hunter in Newcastle on Tyne. The ship left the Tyne ahead of schedule and sailed immediately on her sea trials before being dispatched to the South Atlantic.

But it was the Harrier's maintainability, serviceability and effectiveness—with Blue Fox radar and Sidewinder AAMs—as a fighter that won the air battle and brought the aircraft into the news headlines and to the forefront of public notice.

Their performance throughout the conflict was summed up by Defence Minister Nott in the House of Commons: 'Twenty-eight of our 32 Sea Harriers were deployed to the area and they achieved at least 28 kills without a single loss in air-to-air combat. There were in excess of 2000 operational sorties from the carriers and one of the most remarkable features of the whole operation was the 90 per cent availability of all aircraft embarked.' That is praise indeed, and every word of it fully justified.

The Hercules

'The only replacement for a Hercules is another Hercules.' Lockheed's load-carrying C-130 Hercules has replaced the Dakota as the world's toughest and best-loved aircraft. Both sides used the 'Herc' in the Falklands war, and this is the aircraft's story.

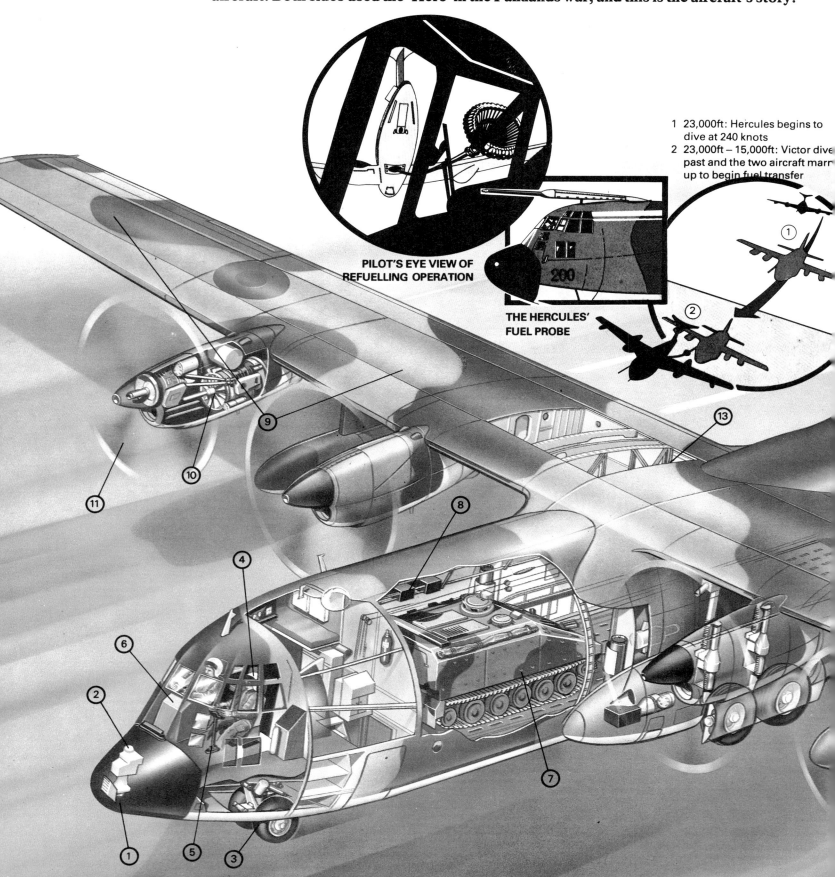

PILOT'S EYE VIEW OF REFUELLING OPERATION

THE HERCULES' FUEL PROBE

1 23,000ft: Hercules begins to dive at 240 knots
2 23,000ft – 15,000ft: Victor dive past and the two aircraft marr up to begin fuel transfer

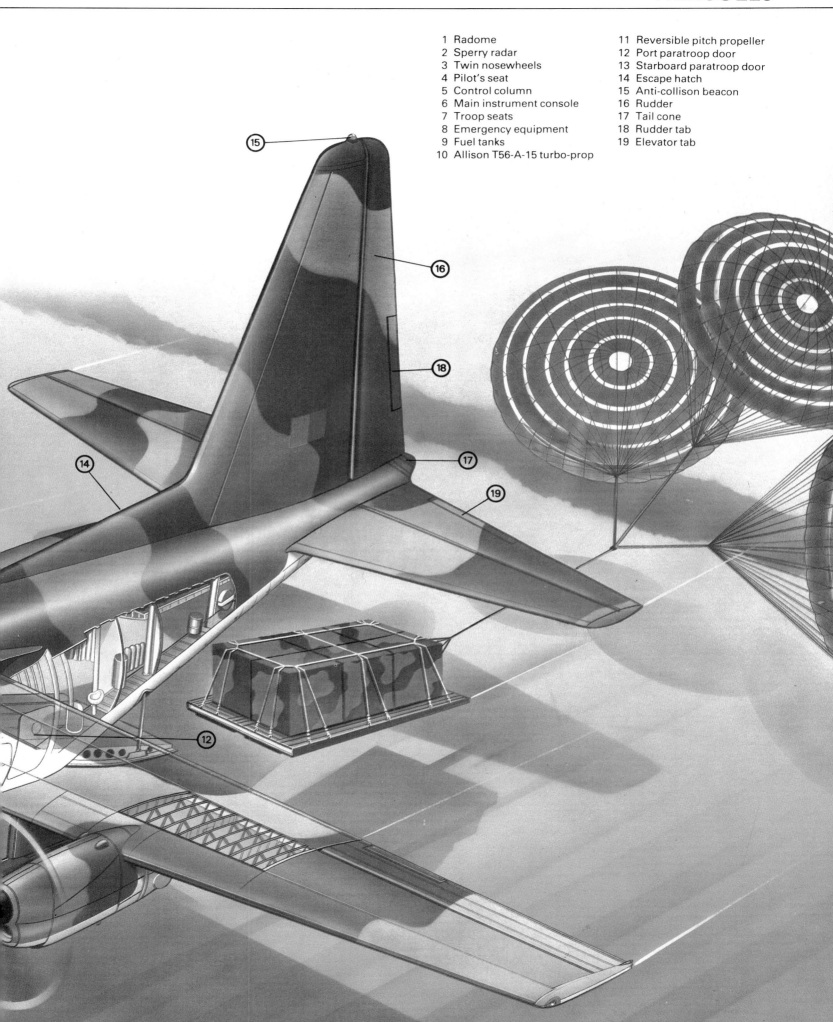

1 Radome
2 Sperry radar
3 Twin nosewheels
4 Pilot's seat
5 Control column
6 Main instrument console
7 Troop seats
8 Emergency equipment
9 Fuel tanks
10 Allison T56-A-15 turbo-prop
11 Reversible pitch propeller
12 Port paratroop door
13 Starboard paratroop door
14 Escape hatch
15 Anti-collison beacon
16 Rudder
17 Tail cone
18 Rudder tab
19 Elevator tab

Lockheed C-130 Hercules

AS WE APPROACH the mid-80s the classic 'Herky Bird' is in the forefront of military transport operations: it may lack the speed and payload capacity of later jet transports, but its 'go anywhere, do anything' capability gives it a major edge over its rivals in terms of tactical utility. Yet in 1984 the type will have been in production for 30 years, providing ample proof of the adage that the only replacement for a *Hercules* is another *Hercules*.

The origins of the *Hercules* go back to 1951, when the US Air Force issued a requirement for a transport to serve with the Tactical Air Command; the winning design had to be able to use rough airstrips, and be capable of carrying 25,000lb of freight, or alternatively a load of 92 infantry or 64 paratroops. Designs from Boeing, Douglas and Fairchild were beaten by the Lockheed Model 82 submission, if which two YC-130 prototypes were ordered. These appeared in 1954, and it was immediately apparent that the world's first truly effective military transport had been born. Earlier aircraft had possessed useful individual features, but in the *Hercules* all the factors necessary for an effective military freighter were combined in a single airframe; a capacious hold (length 41ft 5in, width 10ft, height 9ft) with its floor at truck-bed height for the easy loading of items brought in by lorry; a hydraulically operated ventral ramp/door for the straight-in loading of bulky items or vehicles; the ability to open this door in flight to despatch paratroops or paradrop loads in optimum conditions—cabin pressurization; full span integral tankage for long range; the reliability of turboprop

power; excellent fields of vision for the crew; and STOL field performance and fighter-like manoeuvrability. These last four factors, combined with the design's wide-track blister-mounted landing gear and large tyre area, permitted this big aircraft to operate into and out of primitive airstrips that would otherwise have been impossible even for smaller transports. Particularly important were the combination of the Allison T56 turboprop (four of which were located in slim nacelles projecting forward of the wing leading edge) and 4372 gallons of fuel in integral tanks between the two main spars over the full span, and the provision of a fuselage floor made up of integrally machined panels for lightness and great strength. Durability was given to the type by the use of a new aluminium alloy, and the type also pioneered the extensive use of titanium and bonded metal structures for immense structural integrity.

The USAF were delighted with the *Hercules*, and soon ordered the type into large-scale production. Flight trials revealed a few problems, and the first production aircraft (C-130A series) started to appear from 1955 with few alterations apart from a strengthened rear fuselage and vertical tail (the latter with a squared-off tip), a stronger nosewheel leg, different weather radar in the odd-shaped radome that gives the *Hercules* its distinctive nose, provision for eight 1000lb RATO units and two underwing pylons for

Far left The most widely used
military transport in the
world, the *Hercules* can
discharge heavy loads with
surprising ease. **Left**
Equipped with an air
refuelling probe, this *Hercules*
can take part in long-range
refuelling exercises. **Right** The
aircraft was used by both
sides: a Mercedes jeep is
unloaded from an Argentine
Hercules. It is capable of taking
considerably heavier loads, for
example armoured personnel
carriers. **Below** A US *Hercules*
refuels Skyhawks.

375 gallon auxiliary tanks. The first operational unit
was the 463rd Troop Carrier Wing. It soon enthused
about its new aircraft, which displayed such remarkable capabilities as reaching the stall position
and then flying in that attitude when the throttles
were slammed open.

This versatility meant that the *Hercules* was soon
pressed into a host of alternative roles, made more
than possible by the extra 500 hp provided by the
C-130A's T56-A-9 turboprops. Some 219 C-130As
were built for the USAF, while 12 went to the Royal
Australian Air Force and 32 to the South Vietnamese
air force. Then came a proliferation programme, and
the *Hercules* became a true multi-purpose type.

Some 66 aircraft equivalent to the C-130H—the
latest military variant of the *Hercules*—were produced by Lockheed under USAF contract as C-130K
transports for the RAF, which designated the type
Hercules CMK1. These aircraft were used to
enormous effect in the campaign to recover the
Falklands, staging through Wideawake airfield on
Ascension Island to fly on into the South Atlantic.
Paradrop loads were delivered before a strip on the
islands became available, the *Hercules* transports
involved having to refuel in flight on the legs from
and to Ascension either from Victor KMc2 tankers or
from specially converted *Hercules* tankers. The use
of the former presented fewer logistic problems, but
the incompatibility of the two types' speeds meant
that the so-called 'tobogganing' system had to be
used. Despite considerable difficulties, the *Hercules*
made an enormous contribution to the British success in the Islands, and is still playing the key role in
keeping the garrison supplied and reinforced. It
should be noted that the Argentines made profitable
use of the *Hercules*, this being the only type with the
range and STOL performance to fly from the mainland to Port Stanley and then land clandestinely on
the damaged airstrip.

The *Hercules* tanker concept stretches back some
time, the US forces having operated the *Hercules* in
this form for nearly two decades. The most important US *Hercules* tanker is the KC-130F, a version
of the C-130B for the US Marine Corps. Some 45 of
this model were built, proving especially useful for
the flight-refuelling of other *Hercules* and propeller-
driven aircraft, and also for the support of Sikorsky
HH-3E armed rescue helicopters. A more advanced
tanker is the KC-130R version of the C-130R version
of the C-130H series. Fourteen of these have been
acquired by the US Marine Corps.

Further developments of the *Hercules* are in-
evitable, and Lockheed has even proposed the type
as an aerial minelayer. The order are still coming in
and by 1982 more than 1600 *Hercules* had been
delivered to more than 50 countries.

HMS Invincible: saved by the bell

Before her hull ever touched the water she had sailed through 15 years of vicissitude. Her design had been changed, her function modified, her role re-defined. And when she was needed, she had almost been sold to Australia.

Above *Invincible's* emblem.
Below Restrained by about 1500 tons of chains hanging from her sides and dragging behind along the slipway, *Invincible* is launched at her birthplace, Vickers Shipbuilding and Engineering, Barrow-in-Furness. She had to be restrained because there is only a limited area of deep water off this slipway.
Bottom *Invincible* in the fitting-out stage. The 'ski jump' (invented in 1973 by Lt-Commander Douglas Taylor and developed at the Royal Aircraft Establishment, Bedford) has yet to be fitted.

HMS *Invincible* steamed out of Port William, the outer harbour of Port Stanley in the Falkland Islands, to do battle with a German squadron commanded by Vice-Admiral Graf von Spee. The date was 8 December 1914 and the *Invincible* in question was the fifth ship of that name to serve in the Royal Navy. By employing her main armament of 12in guns she and her consort, HMS *Inflexible*, along with other ships of a hastily assembled British force, destroyed the German squadron except for one ship. Soon after this action *Invincible* rejoined the Grand Fleet in home waters, from which she had been detached at short notice to repel the threat to the Falkland Islands.

Some 65 years later, another HMS *Invincible*, the sixth ship of that name, was again engaging targets in the vicinity of Port Stanley using her main armaments—this time Harriers. Her service with the Royal Navy was about to be terminated, however. The British Government intended to sell her to the Australians. She had been dispatched at very short notice to join a hastily assembled force, this time to liberate the Falkland Islands.

But the name 'Invincible' does not only link the present vessel with a great tradition of fighting ships. Her revolutionary design has also lent her famous name to a new class of carrier. HMS *Illustrious* and HMS *Ark Royal* are both warships of the *Invincible* class. The key elements of their design are the short through deck with a 'ski-jump' on the forward end and a 'scissors' lift which enables aircraft to load on three sides.

The through deck

Invincible's 550 x 42ft through deck evolved from the idea of putting an after flight deck on a cruiser hull. This would be used by Harriers for vertical take-off and landing—the operation of such an after deck was demonstrated on the cruiser HMS *Blake*. But it soon became evident that if a through deck was developed, the Harrier's weapon and fuel-carrying capability could be greatly improved. And a 'ski-jump' or ramp of around seven degrees on the forward end would give the Harrier greater lift on a short rolling take-off. A through deck would also allow for an enlarged internal hangar space, offering better carriage, maintenance and servicing facilities.

Invincible's standard displacement is officially given as 16,000 tons (19,000 tons fully loaded). Her length is 677ft, her beam 105ft and she is powered by four Rolls Royce Olympus TM3B Marine gas turbine engines. Jane's *Fighting Ships* lists her shaft horse power as 112,000 for two shafts, giving her

a maximum speed of 28 knots, and she can go 5000 miles at 18 knots without refuelling. It is said that *Invincible* is the largest warship in the world to be propelled by gas turbines and that she has the largest propellers ever fitted to a ship in the Royal Navy. Her electrical power is derived from eight RP200 1.75MW diesel generators—another Royal Navy record.

Invincible's armament consists of one twin Sea Dart missile launcher. But since the Falklands war, point defence weapons—including the Vulcan-Phalanx gun system—have been added. Her original aircraft complement was listed as five Sea Harriers and nine Sea King helicopters. These were provided by the Fleet Air Arm's 801 and 820 Squadrons. But during the Falklands campaign, she carried eight Sea Harriers from 801 and 899 Squadrons and 15 Sea Kings, including those from 820 Squadron. In May this squadron alone flew 1590 hours, which is equivalent to having two aircraft airborne 24 hours a day throughout the month.

In addition to the anti-submarine capability provided by the Sea Kings, *Invincible* has a Graseby Type 184 sonar mounted on her hull. Her radar outfit includes a Type 1022 for surveillance, a Type 992R for search, two Type 2016 for navigation and a Type 909 for fire control. Her complement is 107 officers, 114 chief petty officers and 560 ratings. This includes the Air Department, comprising 10 officers and 73 ratings and the squadrons, when embarked, of 54 officers and 218 men.

Invincible is fully air-conditioned. She claims to have the biggest air-conditioning system ever fitted in a ship of the Royal Navy.

Carriers cancelled

In many ways *Invincible* was the ship that was almost never built. In 1966 the Labour government cancelled plans to build three new aircraft carriers, drawn up in 1962, and announced that the Royal Navy's fixed wing flying would die out. They accepted the RAF's argument that all air power needed for future maritime operations could be provided by land-based aircraft. For operations outside the NATO area, an island-based strategy was advanced as the alternative to the carrier task force.

As early as 1962 the Air Staff had suggested that ships of around 18,000 tons could be constructed from which helicopters and vertical take-off aeroplanes, then on the drawing board, could operate. These ships would also carry Royal Marine Commandos, bringing them into line with the 'commando carrier' concept which had been pursued by the Royal Navy since the Suez operation.

In 1966, however, the words 'aircraft carrier' were taboo, but all hopes of maintaining air power in the Fleet were not dead. Even the Defence White Paper of 1966 included some reference to the continued fitting out of Tiger Class cruisers to carry helicopters, and successors to such ships were planned. These were to be command and communication ships, it was said. At the same time trials were being carried out on the deck of the light fleet carrier *Bulwark* of the new P1127 vertical short take-off and landing (VSTOL) aircraft.

In a parliamentary debate in 1967 it was suggested that a 'Harrier carrier' should be constructed using a large tanker hull. And in 1969 preparatory work was going ahead on a 'new class of cruiser'. By then the phrase that was being used in official circles was 'Through Deck Cruiser'. This avoided the embarrassing admission that the carrier decision had been rescinded, but it is difficult to see who was deceiving whom by such a transparent trick of phraseology.

The Navy minister spoke of uprating the Harriers' engines and the possibility of operating them from ships. And in 1970 matters were carried a stage further when it was proposed that successors to the Tiger Class cruiser should have the capability of operating VSTOL aircraft, subject to the results of a cost-effectiveness study.

In 1973 a contract was placed with Vickers for 'the first of a new class of cruiser'. The ship was laid down in 1973, and in 1975 she was referred to as HMS *Invincible*, 'the first of a new class of

A serene moment for *Invincible;* her journey south was, by contrast, rather trying. One of her engines became unserviceable, halving her speed until a mid-Atlantic engine change.

INVINCIBLE

THE 'SKI JUMP' TAKE-OFF (Below)

1 When using the 'ski jump', the Harrier performs a short forward roll with engine thrust directed aft. Reaching the ramp's threshold, the plane is doing 60 knots. As it climbs the ramp, the thrust is directed downwards at about 50°. This gives added lift and reduced forward 'push'.

2 On leaving the ramp the Harrier is making about 80 knots. The forward thrust component of the engine accelerates it at about 4 knots/sec while the downward thrust keeps it airborne—airspeed is still too slow for conventional flight.

3 Only when forward speed has reached 110 knots does the jet begin to 'fly', and the downward thrust can be adjusted in favour of forward speed.

4 Just 15 seconds after leaving the ramp airspeed is 140 knots.

KEY AREAS (Right)

 1 Ship's crane
 2 Junior rates' accommodation
 3 Workshop/hangar area
 4 Forward aircraft lift
 5 Flying control office
 6 Life rafts
 7 Reception area
 8 Downtakes, forward engine room
 9 Ship's launch
10 Aft engine room
11 Aft engine room
12 Walkway
13 Officers' accommodation located in this area

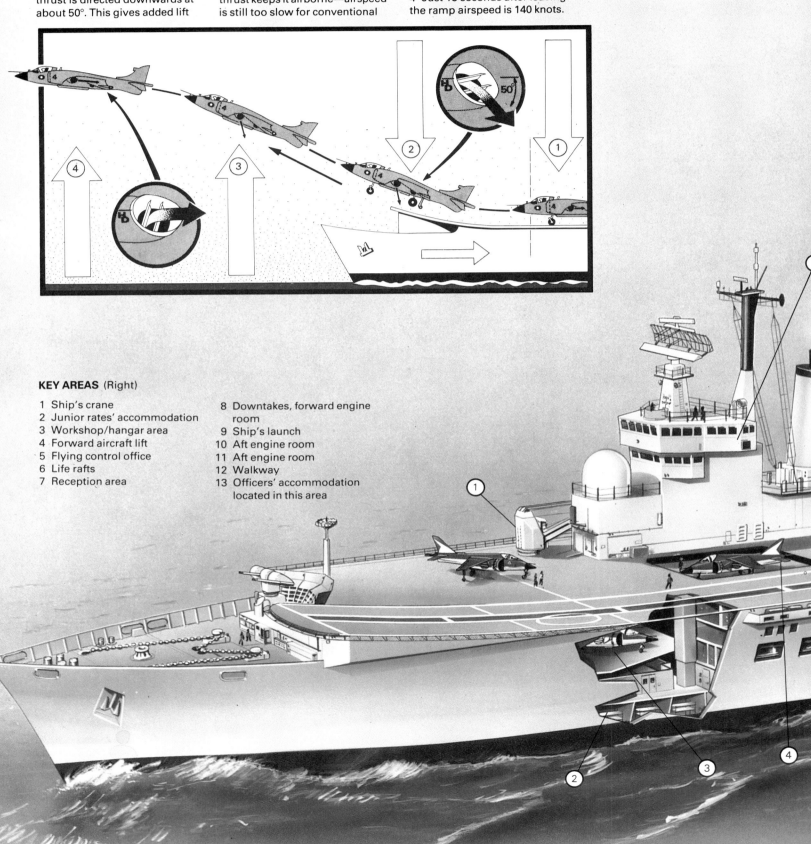

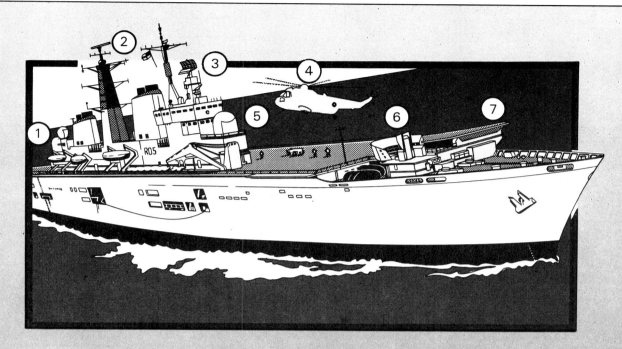

**PRINCIPAL WEAPONS
AND SENSORS** (Left)

1 Aft Type 909 radar (G/H-band)
   for Sea Dart fire control
2 Type 1006 navigation radar
   (X-band)
3 Type 922R surveillance radar
4 Sea King helicopter
5 Forward Type 909 radar
6 Twin Sea Dart launchers
7 'Ski jump' take-off ramp
   inclined at 7°

Below All 678 feet of *Invincible* steam out of Portsmouth. **Bottom** One of *Invincible*'s damage control centres, with a sectional diagram of the ship. After an attack, damage reports are promptly relayed to the control centre.

anti-submarine cruisers'. A second order came and in 1977 it was explained in the Defence Estimates that the prime role of these anti-submarine cruisers was 'to deploy the Sea King anti-submarine warfare helicopter. The cruiser will also provide a command and control function for maritime and task forces and will contribute to area air defence with its Sea Dart and Sea Harrier maritime VSTOL aircraft.' Jane's *Fighting Ships* noted that an amendment to the specification at this stage incorporated the requirement to transport and land commandos. *Invincible*, as she is now, was finally taking shape.

Cruiser becomes carrier

This and other design changes, combined with labour disputes, delayed the construction. HMS *Invincible* was eventually launched on 3 May 1977, the 230th anniversary of the capture of the French *L'Invincible* which became the first HMS *Invincible* in August 1747. The sixth *Invincible* still had to wait another three years before she officially became a carrier. It wasn't until the 1980 Statement on Defence that she was described as 'the first of the class of anti-submarine warfare carriers'. In March of that year she was received into the Royal Navy and in July was commissioned. The overall cost of her construction was £175 million.

Invincible became fully operational in the summer of 1981 and took part in the extensive Allied exercises, Exercise Ocean Venture 81 and Exercise Ocean Safari 81. Despite the designation of this new class of carriers as 'command and control' ships, neither on these exercises nor in the Falklands did *Invincible* command.

Her role was supposed to be that of leader of an anti-submarine warfare group in which her aircraft and missile system would contribute to the air defence and surface support for the group, not supply air cover for an amphibious landing. As the 1966 Defence Review, which killed the old carriers and mooted the 'new class of cruisers', put it: ' . . . only one type of operation exists for which carriers and carrier-borne aircraft would be indispensable, that is landing, or withdrawal, of troops against sophisticated opposition outside the range of land-based aircover. It is only realistic to recognize that we, unaided by allies, could not expect to undertake operations of this character in the 1970s—even if we could afford a larger carrier force.' From that review sprung the idea that any future amphibious landing would be carried out by invitation only, or that British troops would have to await the red carpet treatment in such circumstances.

Amphibious assault

In spite of the official view, the art of amphibious assault remained alive and well. On 21 May one Brigade Group was put ashore without a single casualty, after an approach passage of more than 7000 miles. A second Brigade Group was landed at some cost in face of fierce air opposition and the whole force was supported logistically and with air cover and gunfire of all kinds. On that basis, HMS *Invincible* has justified her place in the Defence Estimates over the past ten years and the twinkle in the eyes of naval planners of two decades.

But at the time of the Falklands landings, *Invincible* was still under sentence. In June 1981 a parliamentary statement called the 'United Kingdom Defence Programme: The Way Forward' said: 'The new carrier *Ark Royal* will be completed as planned but we intend to keep in service in the long term only two of the three ships of this class.'

HMS *Illustrious*, the second of the new class of carrier, was not expected to have been commissioned until early 1983. But she was brought forward to come in to service earlier as a much-needed reserve. And the Defence Statement laid before Parliament in July 1982 expected her 'to join the Fleet later in the year and HMS *Invincible* to be handed over to Australia next year'.

But the demands of the Falklands campaign demonstrated how two carriers, *Hermes* and *Invincible*, were not enough, even when both were, by happy coincidence, fully operational. Mechanical breakdown of either ship between 1 April and 14 June 1982 would have had the most serious consequences, and it is now recognized *politically* that two carriers are not sufficient.

Invincible was finally given her reprieve in December 1982. In a Defence Statement entitled 'The Falklands Campaign: The Lessons', it was stated: 'We have already announced that, following our experience in the Falklands, we intend that two carriers should be available for deployment at short notice. *Hermes* will go out of service when *Ark Royal* is commissioned. To ensure this, a third carrier will be maintained in refit or reserve and we shall not proceed with the sale of HMS *Invincible*.

Rapier: the plane killer

Rapier has been one of Britain's major air defence weapons for several years. Until the Falklands war, however, it had never fired a shot in anger. Once it did, it created a reputation for itself that is the envy of other anti-aircraft missile manufacturers.

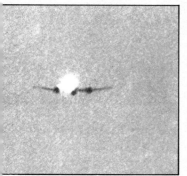

Above Rapier on test. It had proved itself on the firing range. How would it do in war? **Below** The answer came soon—a missile streaks away from its launcher in pursuit of an Argentine aircraft.

UGLY SUSPICIONS that the Task Force was extremely vulnerable to air attack were aroused by the sinking of HMS *Sheffield* on 4 May, and confirmed on 21 May as 3 Commando Brigade established itself at San Carlos.

That first day defending the beach-head had seen the Royal Navy lose HMS *Ardent*, while *Argonaut* and *Antrim* were badly damaged by unexploded bombs and *Brilliant* and *Broadsword* had been strafed by cannonfire. The ships' missile systems were not working well so close to land and the commander aboard *Hermes* was disappointed that the main land based SAM (surface to air missile) system, Rapier, had only been fired ten times to achieve a mere three kills. The British faced defeat if the Argentines' bombs started to explode on contact unless Rapier's performance improved. Over the next three days the Rapier crews sharpened their weapon performance considerably, to end the war credited with 14 enemy aircraft destroyed, with another six probably destroyed. It was a valuable performance that proved that Rapier is a SAM of matchless flexibility and effectiveness in battle conditions.

Versatile, mobile, but above all simple to operate, the Rapier might almost have been designed to provide cover in the Falklands campaign. In its mobile form the entire system is designed to be carried in a converted RCM748 armoured, tracked vehicle of exceptional agility, but this had not been deployed with British forces in May 1982. The Task Force therefore had to make do with the Land Rover-towed version. Wheeled vehicles were useless in the Falklands, but Rapier was easily transported by helicopter, which gave it a great advantage over the Argentine, internationally-manufactured Roland system on hefty trucks.

The positions for Rapier's deployment at San Carlos had been chosen by computer back in Britain. This was the point that caused most mis-judgement of Rapier's performance: it was to be used to protect the landing itself—not units of the fleet standing out in Falklands Water some distance from the shore. Although it achieves a good performance against high altitude targets, it has an exceptional one against highly-manoeuvrable, low flying targets of the sort presented by the Argentine Airforce.

Because so many of the Argentine attacks were directed at the ships of the Task Force rather than its land units, the British SAMs were confronted with some unusual targets. There are not many systems that could cope with the task of firing downwards into mist-shrouded valleys, if only because the land would cause such radar clutter. In addition to this, the Argentine aircraft came in very low.

'Blindfire' system

Modern Rapier units use the 'Blindfire' radar system, which has an all-weather, day and night capability, but the 12 unit battery which initially defended the San Carlos landing did not have 'Blindfire' and relied on optical target tracking. When this method is employed the target is first found on the acquisition radar, which interrogates it to find out whether it is a friendly aircraft or an enemy one. The aimer responds to radar contact with an enemy aircraft by acquiring it in a wide-angled sight before turning to a narrow-angled sight and waiting until a computer tells him that the target is within range before firing. The launcher is under command to align itself with the sight. When the missile flies into view a television camera tracker detects flares in the missile's tail and measures any deviation from the line of sight, which is corrected by radio command. The missile has such a high chance of hitting the target that its warhead is fitted with a contact fuse rather than a proximity fuse. If the aimer keeps his sight on the target the tracker will steer the missile home.

In the action at San Carlos there were unusual difficulties with the radar. The ships' radars were on a similar frequency, which provided interference, and the constant helicopter movements sent out a maddening stream of friendly aircraft pulses to the interrogatory radar. In the end, some operators ignored the acquisition radar to line up with their targets by native wit. The adoption of this technique underlines the system's flexibility, just as a growing rate of kills demonstrated its effectiveness.

Top The brains of Rapier: the command transmitter (the small dish antenna) with the guidance computer in the large dome behind. The small size of the missile is apparent here, but Rapier can afford to be small: its accuracy virtually ensures a hit and so the explosive charge and the fuse assembly can be smaller than in a missile with a proximity fuse. Above The optical sight unit is far smaller despite containing two types of lens.

ANATOMY OF THE RAPIER (Below)

1 Hot gas actuators
2 Tracking flares
3 Control surfaces
4 Control section
5 Cableform ducts
6 Launching foot
7 Wings
8 Motor
9 Electronics and autopilot
10 Warhead, fuse, plus safety and arming unit

Below Setting up a Rapier launcher. Problems created by a long sea voyage and the necessarily rough handling they suffered—on top of radar interference from friendly ships—meant that setting up was a lengthy and frustrating business. The effort was invariably worthwhile.

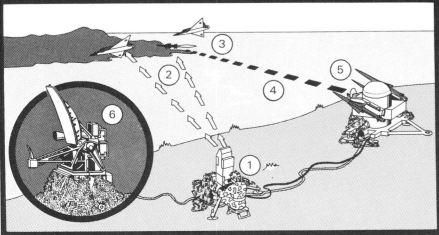

RAPIER GUIDANCE SYSTEM (Left)

1 Optical sight unit
2 Line of sight
3 Flare for tracking
4 Command link
5 Command transmitter computer

EXOCET

1 Homing head
2 Guidance systems
3 Logic frame
4 Fragmentation casing
5 Warhead charge
6 Sustainer motor
7 Booster charge
8 Steering and power supply

Right Inside the Exocet. The sophisticated guidance system includes a radio altimeter, which provides the information that allows the missile to skim low over the sea. The warhead charge is small relative to the overall size of the weapon, but it packs a mighty punch—a 364 lb Hexolite charge. **Below** The MM40 Exocet can be launched either from ship or land—here from a coastal battery firing vehicle.

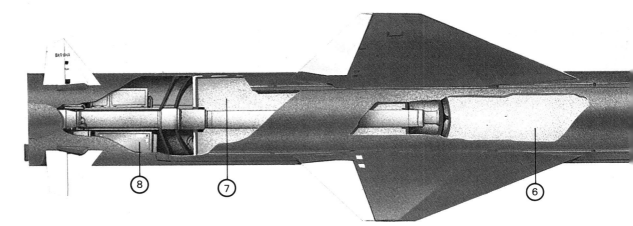

HOW CHAFF WORKS (Right)

1 The RN Corvus chaff launcher.
2 Clouds of chaff show up as two large metallic objects beside the target on the missile's radar (shown in diagram as inset 'bubble').
3 To the missile the chaff clouds are a more 'attractive' target than the ship, so it steers away from the ship towards the chaff.

The Exocet: deadly but beatable

The name of this sea-skimming missile is indelibly stamped on the British consciousness, if not the world's. The Exocet did, of course, wreak considerable havoc on the Task Force. But did it win a reputation far greater than its performance deserved?

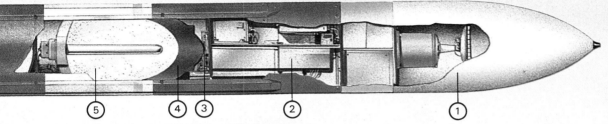

⑤ ④ ③ ② ①

Right A Super Etendard shown carrying an AM39 Exocet missile. This is the weapon that destroyed *Sheffield*. **Below** An MM40 Exocet launched during French Navy trials. Within 2.5 seconds the missile will cruise at 700mph.

IN THE WAKE of *Sheffield*'s loss there was anger in Britain that this severe blow had been suffered at the hands of a French weapon fired from a French-built aircraft, but around the world the telex and cable links to Aérospatiale buzzed with enquiries for the supply of this powerful missile. The actual performance of Exocet, however, was unimpressive. Of the two rounds that struck Task Force warships, one failed to explode and the other gave only a very small detonation that left the ship almost undamaged. Far more impressive was the small British helicopter-launched Sea Skua, which was still in the development stage in April 1982. Hurriedly cleared for service use, seven rounds were fired, all in atrocious blizzard conditions. All seven struck their targets and all seven warheads detonated.

Origin of the Exocet

On 21 October 1967 Egyptian motor boats of the Soviet Osa class fired SSN2 Styx missiles without even leaving Alexandria harbour, and scored direct hits on the Israeli destroyer *Eilat*, which was completely destroyed. The effect on naval staffs was electrifying. Many countries began to develop their own anti-ship missiles, while others urgently worked on the problem of producing a missile that would destroy the small missile boats. One of the latter was Nord-Aviation, which in 1970 became part of the giant conglomerate called Aérospatiale. The result was MM38 Exocet.

Despite the long experience with turbojet cruise missiles such as MM20 it was decided to make a fresh start. Turbojet propulsion was ruled out on the grounds that such engines take too long to start and run up to full thrust. The French company's publicity material overstates the time needed, but the fact remains that a turbojet finds it hard to launch a missile only two or three seconds after the command has been given, unlike the solid rocket motor. Exocet has a two-stage motor, the first having high thrust for the launch and the second being a low-thrust, long-burn sustainer.

Surprisingly, Exocet was given a cruciform-wing configuration, one advantage of which is that there is no need to bank in turns. This gives faster response to commanded changes in heading and may even enable the sea-skimming height to be slightly reduced without risking a wingtip in the sea.

The basic MM38 version has a guidance memory which is fed with the target's range and bearing before launch. When the firing button is pressed the missile is ejected from its launcher, and climbs to a height of 100ft. Then it settles into level flight before dropping to its cruise height of 50ft. Within 2.5 seconds it reaches its cruising speed of Mach 0.93 (about 700mph), so approximately 25 miles can be covered in two minutes, during which time typical target vessels cannot move further than about 1¼ miles and probably much less.

Close to the target, the missile activates its radar which searches until it finds the target and locks on. Unlike most earlier anti-ship missiles, Exocet was designed with a two-axis radar which searches the horizontal plane only, holding the sea-skimming height to final impact. This is considered to make the missile more difficult to detect and shoot down than one which climbs near the target and then swoops down from above. The Exocet is extremely ac-

An MM40 Exocet—the sea-launched version—roars into life. The pieces flying off are simply spacers, which guide and protect the missile during its journey through the launcher itself.

curate—virtually 100 per cent certain to lock on its intended victim.

The warhead is substantial, a 364lb Hexolite charge detonated after passage through the outer plating of the target (which in modern warships is nothing like the thick armour of earlier vessels). Against a missile boat such a warhead would be catastrophic, but it also has a powerful effect against larger vessels. This was demonstrated by the RN's firing of the 101st Exocet off the production line against an old destroyer, which would have been completely knocked out had it been in action.

The Exocet in action

MM38 was designed to be fired from a large aluminium box launcher mounted on the deck of a surface warship. In the Falklands, Argentina mounted the same launchers on wheeled chassis for use from shore. Aérospatiale developed further versions of Exocet, including the AM39 air-launched model carried by large helicopters as well as attack aircraft, the improved high-performance MM40 fired from lightweight tubes on ships or 6 x 6 trucks, and the SM39 with folding wings fired from capsules launched underwater from submarine torpedo tubes.

The main version used by the Argentine Navy in 1982 was the air-launched AM39, carried under the right wing of Super Etendard attack aircraft and balanced by a drop tank under the left wing. To use the missile it was necessary for the aircraft to detect the target ship on its Agave radar and transfer the range and bearing to the missile before release. The missile, in its terminal phase of flight, also had to use radar to home on the target.

Exocet's failure to detonate on *Sheffield* may have been due to a fault or to the inexperience of the user. When fired against the container ship *Atlantic Conveyor* on 25 May two missiles struck the ship and at least one did explode, causing fire which became uncontrollable. The last Exocet fired in the Falklands war, from a land launcher at Port Stanley on the penultimate day, was an MM38 of about 25 miles maximum range.

On this occasion the radar illumination of the ship—the DDG *Glamorgan*—again appears to have gone unremarked, but the missile itself was detected by the ship's active radars and the vessel was smartly turned stern-on to present a smaller target. At the limit of its range the Exocet hit the quarter-deck, passed throught the hangar and entered the crowded galley where it exploded with much less than the normal force. Splinters caused 13 deaths and 14 were injured but damage was slight, and a fire on the mess deck was quickly extinguished.

SEA KING

1 Centre console
2 Pedals
3 Control column
4 Co-pilot's seat
5 Instrument panel
6 Pilot's seat
7 Overhead panel
8 Pitot head
9 Accessories
10 Access Door

11 Turbine
12 Crew entry door
13 Turbine exhaust
14 Rotor brake
15 Cabin floor
16 Navigation light
17 Rotor head
18 Rotor blades
19 Rotating blade pivots
20 Main reduction gearbox
21 Oil pump
22 Gearbox mountings

23 Main undercarriage
24 Flotation bag
25 Sponson
26 Search radar scanner
27 Tail Wheel
28 Tail rotor drive
29 Intermediate 1-1 gearbox
30 Tailplane
31 Tail gearbox
32 Anti-torque tail rotor

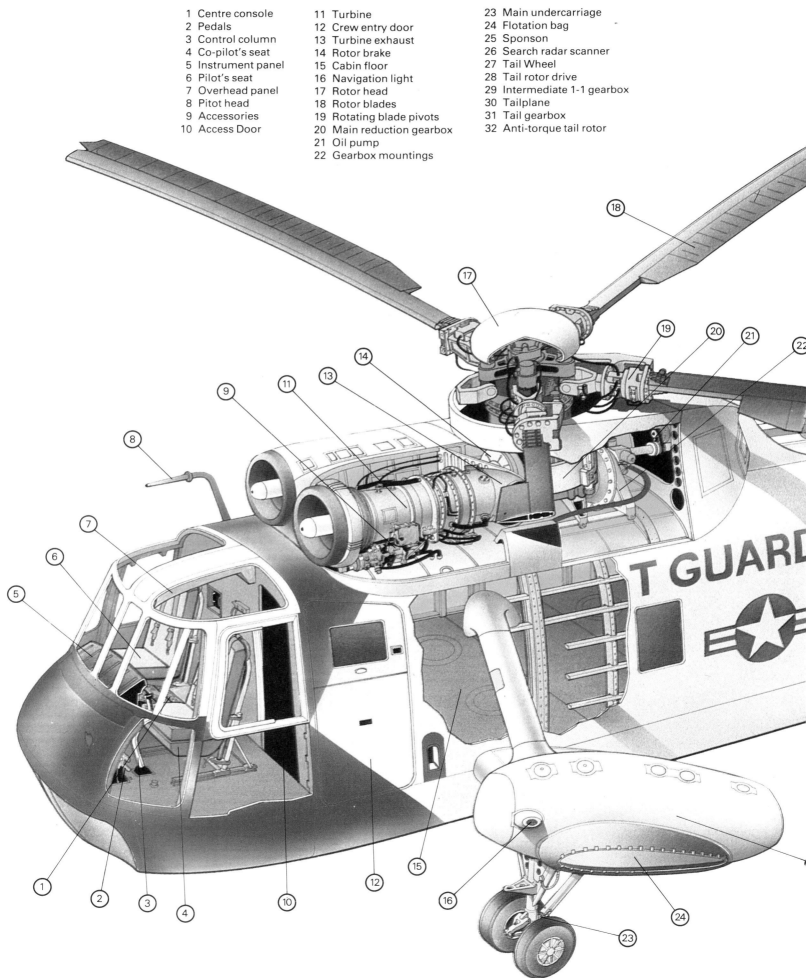

The Sea King

Designed by Sikorsky as an anti-submarine helicopter for the US Navy, the Sea King has been in production for nearly a quarter of a century. In that time it has evolved into an all-purpose helicopter which is both a workforce and weapons platform.

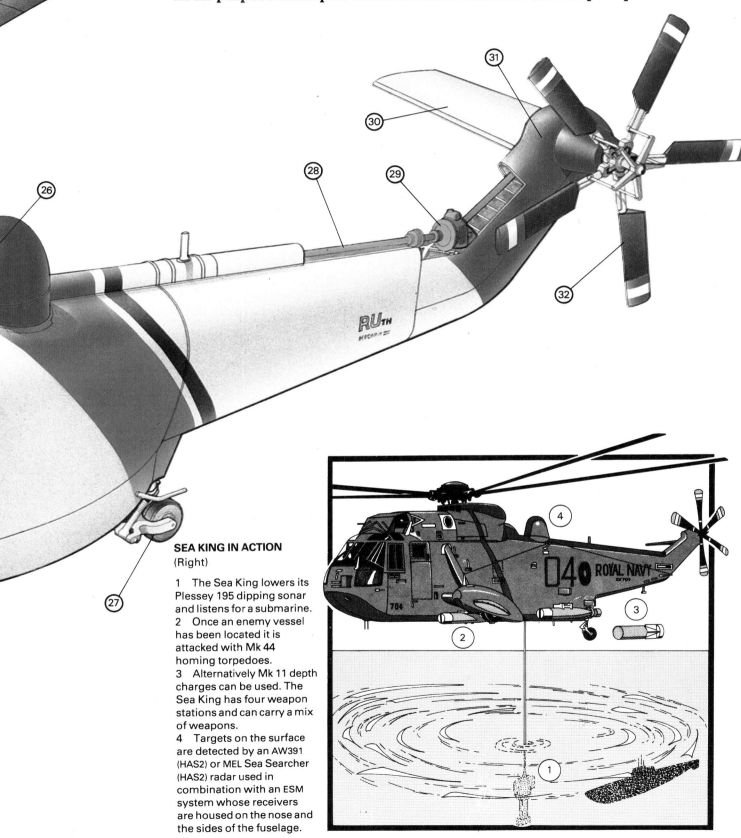

SEA KING IN ACTION
(Right)

1   The Sea King lowers its Plessey 195 dipping sonar and listens for a submarine.
2   Once an enemy vessel has been located it is attacked with Mk 44 homing torpedoes.
3   Alternatively Mk 11 depth charges can be used. The Sea King has four weapon stations and can carry a mix of weapons.
4   Targets on the surface are detected by an AW391 (HAS2) or MEL Sea Searcher (HAS2) radar used in combination with an ESM system whose receivers are housed on the nose and the sides of the fuselage.

SEA KING

THE SEA KING MARKED the arrival of turbine power in military helicopters. This cut the size of the power plant and left more room for the crew and payload. At the same time the Sea King was extremely popular with its crew and passengers because of its engine out 'capability'. This meant that it could fly on one engine if the other failed.

In the late 1940s and early 1950s the rotary-wing aircraft offered great potential in anti-submarine warfare. But the small capacity of the helicopters of that era prevented them carrying effective dunking sonar and weapons. Various expedients were tried. Large Bell HSL-Is and pairs of smaller Sikorsky HSS-Is operated in hunter-killer teams, but their low-powered engines were the chief limiting factor.

The twin turboshafts

The advent of the turboshaft engine offered a powerplant with less vibration, more power and greatly reduced volume and weight. The helicopter was transformed, and the Sea King was designed in 1957.

The aircraft was conceived in the traditional Sikorsky mould. It had a fully articulated main rotor, multi-blade tail rotor and all-metal pod-and-boom fuselage of stressed-skin flush-riveted construction. But it was the engine installation that was novel. Twin turboshafts were located side-by-side above the cabin and drove directly into the main gearbox. Previous massive piston engines had been placed at an oblique angle in the nose with a heavy transmission shaft running through the aircraft.

The first Sea King flew on 11 March 1959 and was able to carry both the advanced AQS-13 dunking sonar and weapons up to a weight of 840lb. The new helicopter could search an area 10 times faster than its predecessor. Development was fast and furious in the 1960s. The incorporation of a watertight boat hull and outrigger-mounted stabilizer sponsons made the basic design even safer. A luxury version was even added to the US presidential flight. Sea Kings were used as utility transports resupplying 'Texas Tower' radar platforms in the Atlantic Ocean, for the movement of equipment into missile bases.

The transport models

This in turn launched the transport model, which was bought by the Royal Danish and Royal Malaysian air forces, and two specialized airliner derivatives—a landplane model with fixed landing gear replacing the amphibious type and a larger model with retractable landing gear and a stretched fuselage, inbuilt sonar equipment replaced by a removable package allowing the amphibious transport version to be used for utility and anti-submarine warfare when needed. But the most important innovation is the LN-66HP radar (with its antenna in a retractable radome under the fuselage) for the detection of incoming anti-ship missiles, a feature that would have been of immense importance for the British Sea Kings operating in the Falklands campaign.

The US Air Force version did not need full amphibious capability, so the stabilizer sponsons were replaced by small stubs located farther aft, allowing the landing gear layout to be altered to a fully retractable tricycle configuration. The fuselage was altered to provide a large rear ramp at the rear

Below One of the major shortcomings of the Sea Kings deployed in the Falklands campaign was their lack of missile detection radar. After the loss of *Sheffield* this was improvized by adding a Searchwater radar with a large radome underneath which swivelled back for landing.

Right The workhouse Sea King can carry up to 5000lb of freight. This means that it can transport light artillery and small vehicles as well as men. In the Falklands the Mk 4 fixed-wheel dedicated transport version, known as the Commando, was used for freight carrying duties.

end of the 'pod' for direct access to the hold; freight or even small vehicles could be loaded, permitting paradrops to front-line units. The tail plane was also much enlarged and given a bracing strut. These aircraft played an important part in Vietnam as utility transports. More important, perhaps, was the Jolly Green Giant rescue helicopter version which had provision for inflight-refuelling and served for the rescue of downed aircrews. The US Navy had its own rescue variant fitted.

The British approach

In Japan Mitsubishi had built two series with only minimal differences from the US aircraft, and in Italy Agusta is also building helicopters closely modelled on American versions. In the UK, the airframe and powerplant of the Westland Sea King are very similar to those of the American aircraft, despite the fact that the powerplant comprises two Rolls-Royce turboshafts, but the avionics and operational equipment are markedly different. This is largely as a result of different tactical thinking. The US tends to think of its Sea King as long-range sensor platforms and destroyers are called in for the kill. The British philosophy is to use the Kings as complete weapon systems which both hunt and kill submarines. In the

British Sea Kings, one of the two cabin crew sonar operators is replaced by a tactical co-ordinator, who controls the operation of the helicopter up to the release of the weapons. These consist of up to four Mk 44 homing torpedos or Mk 11 depth charges. Completing the package is an Ekco AW391 radar with its antenna in a dorsal radome aft of the rotor assembly.

Deliveries of these initial Westland Sea Kings began in the late 1960s, and it was the Mk 5 version which played a most important part in the Falklands campaign. The threat posed by the Sea Kings is thought to have kept the Argentine submarines well out of the way of the Task Force. Operating at far more than their normal rate, and often under the worst weather conditions, the Mk 5 Sea Kings kept up constant vigil while Mk 4s ferried men and equipment between ships and between ship and shore.

The fatal omission

But the one equipment omission that spelled disaster in the Falklands campaign was Airborne Early Warning radar. With a speed impossible in peace time, an improvized AEW aircraft was produced by adding Mk Thorn-EMI Searchwater radar. This has its antenna in a large radome projecting from the starboard side of the fuselage on a rotating mounting. In use, this is swivelled down to the vertical position, otherwise it is turned horizontally to the rear. This extemporized mounting has proved very successful, and it is likely that five Sea Kings will be converted in this fashion. And to replace losses in the Falklands campaign, 12 new Sea Kings are to be built.

Below The Royal Navy use Sea Kings as hunter killers, detecting submarines by sonar and attacking them with Stingray torpedoes or Mk II depth charges.

The Mirage: a plane too far from home

The Mirage family of aircraft are proven in combat and adaptable to several roles: a copy-book export success for their French manufacturers. The Falklands war undermined this well-deserved reputation in a way that wasn't fair on plane or pilot.

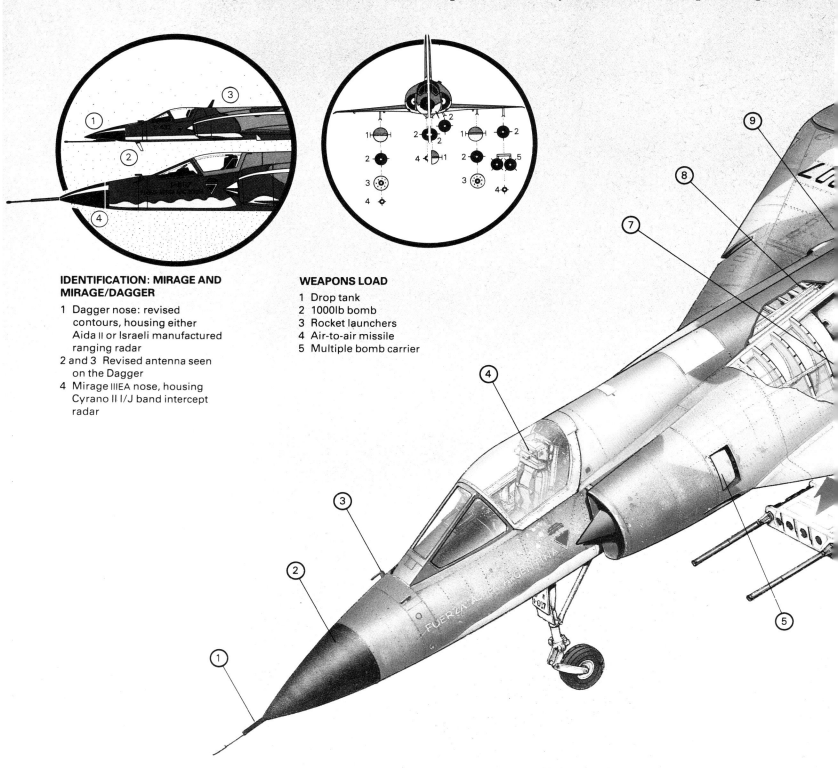

IDENTIFICATION: MIRAGE AND MIRAGE/DAGGER

1 Dagger nose: revised contours, housing either Aida II or Israeli manufactured ranging radar
2 and 3 Revised antenna seen on the Dagger
4 Mirage IIIEA nose, housing Cyrano II I/J band intercept radar

WEAPONS LOAD

1 Drop tank
2 1000lb bomb
3 Rocket launchers
4 Air-to-air missile
5 Multiple bomb carrier

ANATOMY OF A MIRAGE

1 Pitot tube
2 Glass fibre radome
3 Instrument pressure sensors
4 Martin Baker RM4 ejection seat
5 Auxiliary air intake door
6 30mm DEFA cannons
7 Fuselage fuel tanks
8 Dorsal system ductings
9 Cooling system air intakes
10 SNECMA Atar 9c afterburning turbo jet
11 Airbrake
12 Hydraulic undercarriage retracting jack
13 Stores pylon
14 Leading edge notch
15 Rudder hydraulic actuator
16 Glass fibre fin tip aerial
17 Brake parachute fairing
18 Exhaust nozzle shroud
19 Cooling air louvres
20 Variable area exhaust nozzles
21 Jet pipe
22 Engine bay/jet pipe thermal lining
23 Wing integral fuel tank
24 Elevons

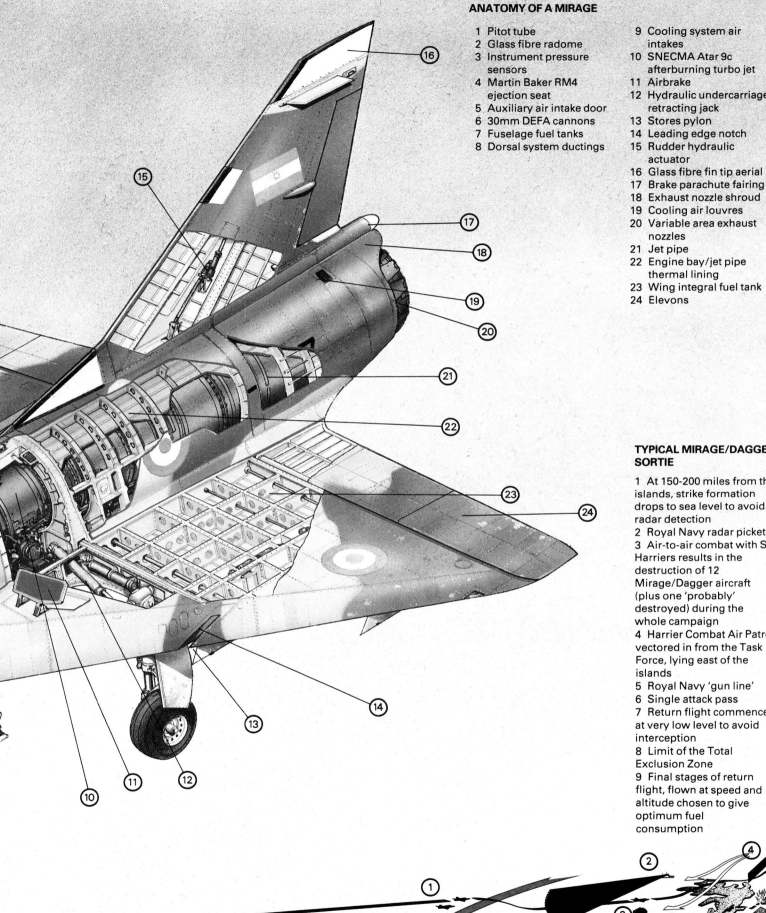

TYPICAL MIRAGE/DAGGER SORTIE

1 At 150-200 miles from the islands, strike formation drops to sea level to avoid radar detection
2 Royal Navy radar picket
3 Air-to-air combat with Sea Harriers results in the destruction of 12 Mirage/Dagger aircraft (plus one 'probably' destroyed) during the whole campaign
4 Harrier Combat Air Patrol vectored in from the Task Force, lying east of the islands
5 Royal Navy 'gun line'
6 Single attack pass
7 Return flight commenced at very low level to avoid interception
8 Limit of the Total Exclusion Zone
9 Final stages of return flight, flown at speed and altitude chosen to give optimum fuel consumption

MIRAGE

ON PAPER, the Argentine Air Force's supersonic Mirage and Mirage/Dagger jets were more than a match for their sub-sonic enemy, the Harrier. Why then did they fare so badly?

The answer is simple: an accident of geography forced them to fight right at the edge of their range. And they had to be used in a role for which they were not procured.

The natural role of the Mirage and Mirage/Dagger is as a fighter-interceptor, engaging in classic air combat at altitudes up to a ceiling approaching 56,000 feet. In the Falklands war they were pressed into service as fighter-bombers. Admittedly, a number of the Argentine Mirages were adapted for the surface-attack role, but this was of little account in the face of the over-riding fuel problem.

It is roughly 400 nautical miles from mainland Argentina to the Falklands. The combat radius for the Argentine Mirage IIIE is 647 nautical miles, and that includes a margin of fuel for combat. On the face of it, this seems to be enough to mount successful raids against the Falklands. But it has to be remembered that the figure is for aircraft in 'clean' configuration. Carrying external fuel tanks, which is vital, creates extra drag and increases fuel consumption; likewise externally mounted bombs. It is estimated that the aircraft had enough fuel for only two to three minutes over the target, in other words, time for a single pass. If a pilot cut in his afterburner to gain the acceleration and speed that might have got him away from trouble, then it is doubtful he would have made it home.

So it was not that the Argentine aircraft and pilots failed to perform well. They simply did not have the opportunity to do so. It must have been depressing for the Argentine pilots that they could not repeat the success enjoyed by their Israeli counterparts, particularly in the 1967 Arab-Israeli war. The Argentines were almost certainly as good as the Israelis.

The Falklands war was indeed a hiccough in the long success story of the Mirage family of fighters, fighter-bombers and reconnaissance aircraft. Their evolution began in 1952 with a French Air Force (*Armée de l'Air*) requirement for a lightweight interceptor to be powered by two small turbojets. This was translated into hardware with the prototype Mirage I, powered by a pair of French-built Viper turbojets. It made its maiden flight in June 1955, and was supersonic in a dive, but only with the addition of a liquid rocket motor did it become supersonic (Mach 1.3) in level flight. The manufacturers, Avions Marcel Dassault (later to merge with Breguet Aviation to become Dassault-Breguet) were not satisfied with this low-powered aircraft, and after looking at a Mirage II, they developed the heavier and larger Mirage III, powered by an Atar 101G turbojet. First flown in November 1956, it went supersonic in level flight on 30 January 1957 without the aid of the SNECMA (Société Nationale d'Etude et de Construction de Moteurs d'Aviation) rocket motor fitted to the aircraft.

Exceeding Mach 2

The *Armée de l'Air* realized the potential of the Mirage III, and the design was refined with a re-designed fuselage to accept the new Atar 9 turbojet and a thinner wing. Ten pre-production Mirage IIIA aircraft were ordered, the first flying in May 1958, and in October that year it became the first European fighter to exceed Mach 2 in level flight. The production version, the Mirage IIIC, was soon ordered into service. Fitted with Cyrano I fire control radar, the main armament was to be either a single Matra R530 air-to-air missile (AAM) or a pair of AIM9 Side-

The IIIE fuselage was lengthened slightly to accommodate improved avionics and navigation equipment, plus the improved Cyrano II radar. Deliveries of production aircraft began in 1964, and this version was supplied by France to 13 foreign air forces. These included the 21 Mirage IIIEAs and IIIRAs supplied to Argentina between 1972 and 1980. The RAAF (Royal Australian Air Force) also ordered a version of the IIIE, but with a different avionics fit, as the Mirage IIIO, 98 examples of which were procured, plus two pattern aircraft so that the Australians could produce it under licence. A reconnaissance version of the IIIE, with five nose-mounted cameras, was developed as the IIIR.

Israeli experience with the IIICJ, emphasizing the need for a clear-weather ground-attack version of the Mirage III, led to the development of the simplified Mirage 5. This version first flew in May 1967, and Israel ordered and paid for 50 such aircraft. However, change of policy by the French government towards Israel ensured that none was delivered. (The money was later refunded, and the aircraft passed into French service.) This version was armed with twin 30mm cannon and could carry more fuel and ordnance (8820lb), but it did not have the radar and associated avionics installed.

The next variant of the Mirage III series was unveiled in 1975 as the Mirage 50. This took the Mirage 5 airframe and married it to the Atar 9K50 turbojet of the Mirage F1C, with the addition of either the Cyrano IV or Agave radar, a head-up display (HUD) and improved nav/attack systems.

The 'New Generation'

The Mirage family still had not stopped proliferating. It went on to encompass the F1 series, the Mirage 2000 and the Super Mirage 4000. Most recently, Dassault-Breguet have reworked the basic Mirage III, adapting it to modern technology. The wing is refined. There are canard control surfaces: small aerodynamic surfaces mounted on the upper part of the air-intake just aft of the cockpit. There are fly-by-wire controls, giving much better response to the flying controls by means of electrical signals rather than direct, mechanical linkage. There are improved avionics and Electronic Counter Measures (ECM) systems. The aircraft was relaunched as the IIING (New Generation), the prototype first flying on 21 December 1982.

Israeli interest in the Mirage III did not end with the refusal by France to deliver the Mirage 5Js ordered and paid for. The decision was made to manufacture an aircraft generally similar to the Mirage III, but with an American J79 turbojet in place of the Atar 9. This became the Kfir. As an interim step, however, Israel built an aircraft based on the Mirage III/5 airframe, with an Atar 9C turbojet and Israeli electronics and equipment. The first flight of the Nesher, as it was named, is thought to have taken place in September 1969, with deliveries beginning in 1972. Between 1978 and 1979, 26 Neshers were sold to Argentina, where they are known as Daggers. Ten more followed in 1981.

The official British claim for Mirage and Mirage/Dagger kills by all arms of the Task Force is 27 aircraft. The Argentines admit to losing 12 Daggers, with six more too badly damaged for further use. But this does not take into account the number which never reached base on the way home.

Top An Israeli Air Force Kfir: when France refused to supply more Mirages to Israel for political reasons, plans for the Mirage III were 'obtained' by Israel, who then proceeded to make their own similar, and effective, version of the aircraft. It was powered by an American turbojet in place of the French one and sold to Argentina as the Dagger. **Above** A Mirage 2000, nearly the most-up-to-date of the family. Only the IIING (New Generation) Mirages are more recent, with fly-by-wire controls and improved avionics and ECM systems. **Above left** A Mirage III: numerically the bigger 'generation' of Mirage was the III, comprising no less than six different versions, ranging from two-seat trainers to the IIIE. **Left** Mirage attacking *Bedivere*—its bombs failed to explode.

winders. Additionally, a pair of 30mm DEFA cannon, or the rocket motor, could be carried. The Mirage IIIC was successfully exported to Israel as the IIICJ, to South Africa as the IIICZ, and with a Hughes Taran radar and provision for HM55 Falcon AAMs as the IIIS for Switzerland, where it was produced under licence.

The Mirage IIIB is the tandem two-seat operational trainer variant, which first flew in October 1959. With a slightly longer fuselage and the radar deleted, it can operate in the ground-attack role. An improved version, the IIID, was later developed as the operational trainer version of the IIIE and 5 versions of the Mirage.

The Mirage IIIC was taken a step further in the early 1960s when the longer-range intruder version, the IIIE, was developed. The intruder role is essentially one of going behind enemy lines to bomb non-military, but nonetheless strategic, targets such as rail depots. It involves low-level flight over unfamiliar terrain, to escape radar detection, and for this an inertial navigation system, giving a continuous read-out of position in relation to the target, is essential. Air combat does not require this type of 'avionics', instead relying on radar detection of the enemy and a simpler type of navigational system.

Milan: the bunker buster

It was designed to smash holes in tank armour, but the Milan was also useful, not to say deadly, against Argentines dug into hillsides, even when they were protected by a screen of rock. The British made the most of this new weapon all through the war.

THE TRAINING AND EQUIPMENT of British infantry has for the last 15 years been directed largely towards facing one particular type of enemy in open warfare: a large, mobile armoured force whose main battle tanks are exceptionally well protected. To combat this threat, support companies use a highly accurate, French-built missile system with considerable range and hitting power—the *Missile d'Infantrie Leger Anti-char Milan*. During the Falklands campaign, Argentine armoured units were not involved in the fighting, but Milan was nevertheless blooded as a brutally effective bunker-busting weapon.

It has long been acknowledged that anti-tank weapons have a useful secondary role in house-fighting or attacking strongpoints, and the development of certain characteristics in modern systems like Milan have made them doubly effective in that role. As tanks have become tougher to crack, anti-tank power has been increased. Milan's warhead contains 3.19lb of shaped explosive charge designed to burn right through tank armour. When it hits a hillside dugout or sangar (a hollow shielded by some breastwork), its crushing impact is reinforced by a shock wave and sheet of flame which terrifies those it does not kill. Milan's accuracy has been enhanced by a computer-controlled, wire-guided system which, it is claimed, gives it a 98 per cent chance of a direct hit at ranges between 230 and 2000 yards. The laser rangefinders and computer fire controls of the latest tanks give them a lethal accuracy, and infantry anti-

tank weapons must have a low profile and be elusive. Milan meets this requirement because it is launched from the prone position and is capable of being carried and operated by a single soldier. Well-trained infantry can bring it forward even into open ground swept by enemy fire, as they did in the Falklands.

Although Milan had never been used in anger, the British were keen to discover its potential and used it enthusiastically from the very start of the campaign. Several Milan rounds contributed to the softening-up barrage that persuaded the garrison of South Georgia to surrender, and it was taken ashore with the SAS in the Falklands themselves before the regular infantry units made their landing.

An unnamed member of the SAS remarked that the raid on Goose Green included the hardest march or 'hack' that he had ever done with the SAS. Given the weight and munitions needed for the task this was not surprising. Milan is designed for a man to carry, but its launching pad and guidance unit does weigh 36lb and each missile weighs 24lb when packed, a load which even the strongest soldier will find difficult to carry for long periods. As a result, Milan is not used in a section anti-tank role, but in three-man teams in a battalion's support company.

By the standard of support company weapons, however, Milan remains very portable, which was of

Below Milan under test. The low profile offered by the launcher is immediately apparent. The advantage of this is immense: the missile can be fired at any target without the need for the operator to expose himself to fire.
Inset Coming within range of a heavily defended bunker like this could be fatal—but once these bunkers had been located their defenders couldn't stand up to the devastating firepower of the British Milans.

great significance to 2 Para in the advance on Goose Green on the night of 26/27 May. The lack of helicopter and tracked transport meant that the battalion had to carry virtually all its equipment on the long march from Sussex Mountains to Camilla Creek House on its way to the battle.

A complex produce of the computer age, Milan nevertheless has a rugged outward simplicity which makes it swift to load and easy to operate. Each round of ammunition is factory loaded into a sealed tube which also serves as a launcher. To bring it into action this container/launcher tube is clipped to the mechanical and electrical fittings on the command and guidance unit, which basically is a periscopic optical sight combined with an infra-red tracking and guidance system mounted on a tripod.

The proof
The system proved itself at Goose Green on the morning of 28 May. As 2 Para's attack made progress, support from mortars and guns fell away. Mortar ammunition ran out, the frigate *Arrow*, providing naval gun support, was forced to retire for fear of air attack, and the soft ground made it difficult to keep all three 105mm guns of the Royal Artillery in action at once as the gun trails buried themselves further with the recoil from each round. After first light on 28 May, the paratroops' advance was bogged down. The Argentines were provided with unlimited ammunition in strong defensive positions and, as the ammunition and fire support of the British dwindled, 2 Para was pinned down by heavy and accurate machine gun and sniper fire. At this point, Lt-Colonel 'H' Jones ordered the Milan missile teams to come forward down the spine of the isthmus between Darwin and Goose Green to take on the Argentine strongpoints in front of his B Company.

The Milan operators settled to their task about 1400 yards away from the Argentine bunkers. In many ways the distance was ideal because, although Milan can be effective as close as 20 yards, it takes a little time to gather the missile on to the line of sight, and the system is most accurate at between 230 and 2000 yards. As they identified their targets, the operators fixed them in the cross-hairs of the Milan

sights and fired. On firing, a gas-generated booster charge flares for 45 milliseconds, thrusting the missile forward at 68 yards per second and ejecting the spent tube backwards. When the booster charge is exhausted, the missile's two-stage propulsion motor takes over, providing a rapid but decreasing acceleration to achieve a velocity of 180 yards a second.

Milan's velocity is twice that of earlier portable missiles, but it still takes about 10 seconds to cover 1400 yards and during the time of flight the operator has to keep the cross-hairs of the sight on the target for the correct guidance commands to reach the missile. During its flight the flare of the tail of the missile emits an infra-red signature which is picked up by a tracker on the operator mechanism. This enables the system's computer to measure any deviation between the missile's position and the line of sight. If there is any error, the computer automatically generates command signals to the missile through an attached cable which is payed out through a spool on the missile during its flight. The formidable warhead is kept under constant guidance and control until its crush fuse is detonated by impact on the target.

At Goose Green the system fulfilled British expectations, and the Argentine strongpoints wilted under the combined fire of B Company with the machine gun and Milan teams of Support Company. The position was outflanked and then attacked by D Company, which was the signal for the Argentine soldiers to surrender cluster by cluster. The 97 prisoners taken seemed shocked and dazed by the experience.

Goose Green marked the first occasion on which Milan's weight of fire had been directed at clearly defined targets. Before that, it had been used for a demonstration of fire-power designed to overawe and to add its weight to a diversionary raid. In the battle 2 Para lost only 18 men, while some fifty defenders died. Conventional wisdom asserts that attack is three times more costly than defence, and this remarkable reversal of the odds must be explained in part by the power of modern weapons.

Above left A two-man team can assemble the system and fire two missiles within 50 seconds of leaving a vehicle. **Above** The firing post is low and a periscopic sight enables the operator to keep down. **Below** Taking the firing post from its sealed polyester case. **Bottom** The missile is positioned on the firing post in seconds: it need only be pushed forward on the ramp.

MILAN

1 Ignition
2 Missile moves forward in launch tube; tube moves to rear
3 Missile in free flight; launch tube thrown clear
4 Control wire paid out from rear of missile
5 Infra-red flare at rear of missile
6 Maximum range 2000yds
7 Steering commands pass through control wire
8 Line of sight of operator—if target moves missile follows
9 What the operator sees
10 Launch tube and venturi thrown to rear

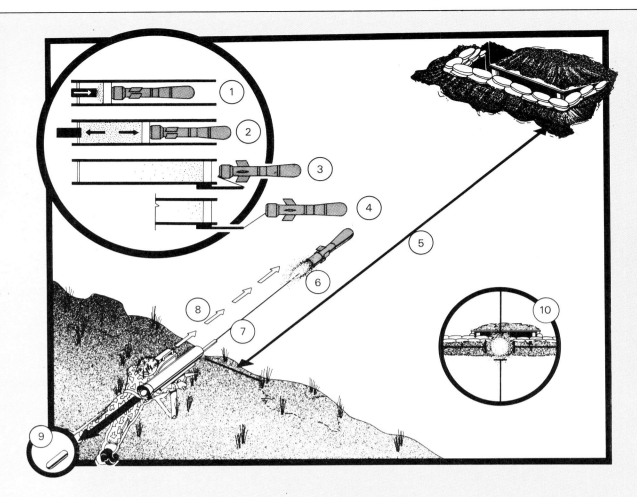

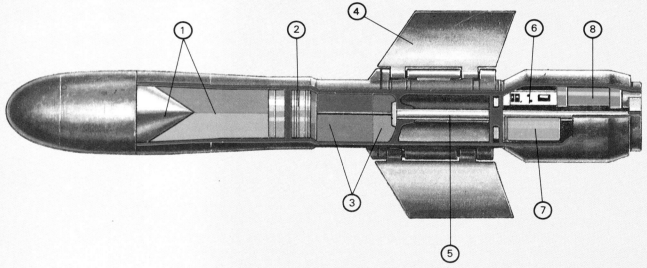

1 Shaped charge warhead
2 Fuse
3 Propellant (two speed)
4 Fin
5 Exhaust discharge tube
6 Decoder
7 Thermal battery
8 Day tracer

Above and right Milan is a very compact, effective weapon. The missile carries a HEAT (high explosive anti-tank) warhead with a shaped explosive charge. This sends a jet of superheated gas and molten metal through anything unlucky enough to get in its way. Milan's only drawback is its comparative weight—it is slightly too heavy to be comfortably manpacked.

Far left Royal Marines practise with Milan in Norway during the winter of 1980. By the time Milan went into action for real, both paras and marines were fully trained in its use, and had spent countless hours on the Milan simulator.

Left One item of equipment which may come in handy later on is a quick-release bracket for mounting the unit on a vehicle

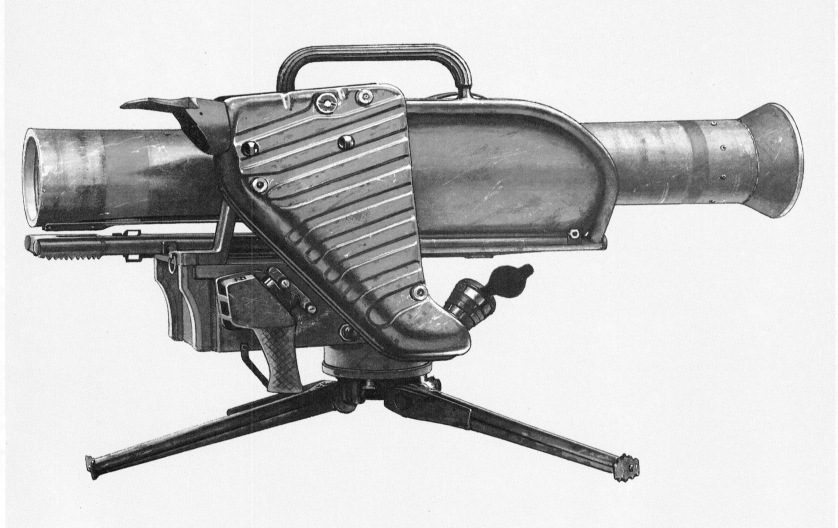

A family of missiles

Seawolf, Sea Dart and Sea Cat bore the brunt of the anti-aircraft work at San Carlos and after. With 20 years between Seawolf and Sea Cat, their performance was remarkably consistent—a credit to the missiles and the men who operated them .

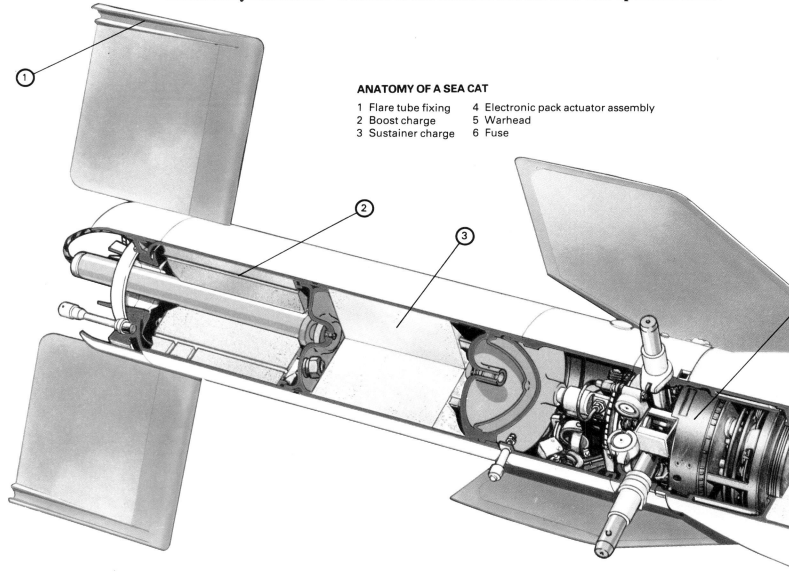

ANATOMY OF A SEA CAT

1 Flare tube fixing
2 Boost charge
3 Sustainer charge
4 Electronic pack actuator assembly
5 Warhead
6 Fuse

SEA DART

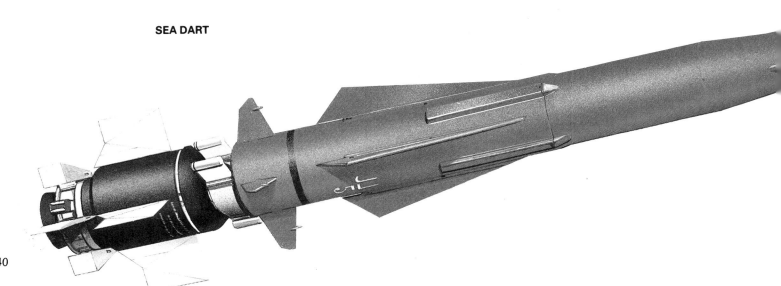

VITAL STATISTICS

1 Sea Dart is effective against high-and low-level aircraft and missiles
2 Sea Dart's range is about 185 miles
3 Sea Cat is a point defence system effective against low and medium level aircraft
4 Sea Cat's range is about 6000 yards
5 Seawolf is effective low- and medium-level aircraft, missiles and shells
6 Seawolf's range is classified but known not to be great

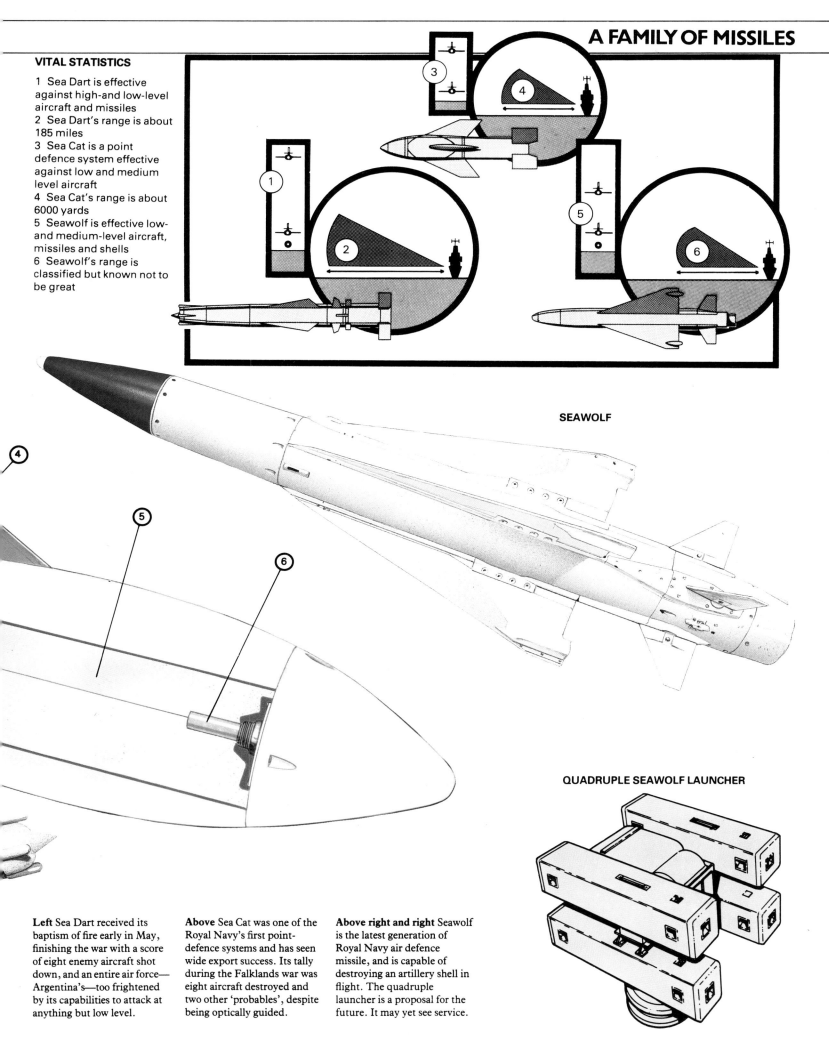

SEAWOLF

QUADRUPLE SEAWOLF LAUNCHER

Left Sea Dart received its baptism of fire early in May, finishing the war with a score of eight enemy aircraft shot down, and an entire air force—Argentina's—too frightened by its capabilities to attack at anything but low level.

Above Sea Cat was one of the Royal Navy's first point-defence systems and has seen wide export success. Its tally during the Falklands war was eight aircraft destroyed and two other 'probables', despite being optically guided.

Above right and right Seawolf is the latest generation of Royal Navy air defence missile, and is capable of destroying an artillery shell in flight. The quadruple launcher is a proposal for the future. It may yet see service.

A FAMILY OF MISSILES

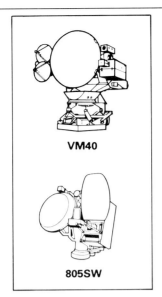

VM40

805SW

Above Two types of Seawolf tracking radar: the upper is the Signaal VM-40 unit used at present, the lower is a new proposal from Ferranti, the 805SW. Both are designed for the Seawolf system and offer resistance to jamming as well as low-level accuracy.
Below Sea Cat is an elderly but effective system that was fitted to several ships, not least the carriers. Sea Cat is credited with several kills, although it was not so effective against some targets as Rapier, Seawolf and Sea Dart. Sea Cat was introduced in 1962 to counter low-flying super-sonic aircraft, so was a direct predecessor of Seawolf.

OF ALL THE BRITISH surface-to-air weapons deployed, it was not only the most modern but also the only one designed as a point-defence system, and was capable of countering the threat posed by sea-skimming missiles like Exocet. After the guidance system was reprogrammed, during the course of the action the missile had greater success in taking on low flying aircraft; it claimed five Argentine jets before hostilities ended.

Its first trials had been encouraging enough. By the time it entered Royal Navy service in 1979 it had already proved capable of hitting a 4.5in shell in flight. But a shell travels on a trajectory considerably higher than a sea-skimming missile and it is in this difficult low cover area that a point-defence system has to excel. In fact, there were only two Type 22 frigates armed with Seawolf available for action off the Falklands. Neither of them came under attack from Exocet, so the system was never put to the test. But the problems of taking on sea-skimmers are well understood and the Seawolf GWS25 system used off the Falklands incorporated high-and low-level surveillance radars as well as an extra TV tracking system for low-level targets. Targets are first located by the surveillance radar, which constantly sweeps round the ship. The incoming target is then probed to find out whether it is friendly or hostile. If this interrogatory radar probe does not receive the proper pulsed radio answering signal sent out by friendly aircraft the target is identified as hostile and assigned to a tracker.

In most cases this tracker will be a radar which establishes a line of sight to the target for the command computer. The system then fires its 176lb 6ft-long Seawolf missile. This is followed by another tracker radar which will establish its line of flight for the computer. The computer can then compare the line of sight and line of flight and radio course corrections to the missile over Seawolf's microwave link.

The Type 910 tracking radar used in the GWS25 system works well up to a very high angle of elevation. Its performance is not so good at low levels. So low-level targets are assigned to a TV tracker and the missile's flight is followed by a second TV tracker. The Falklands war underlined the importance of a point-defence system for all types of ship. *Atlantic Conveyor*, for instance, had no defence against the Exocet that destroyed her, though at 14,946 tons she was big enough to take a GWS25 system. Other smaller ships were endangered by missile attack and it is being urgently considered whether the lightweight Seawolf VM40 system which can be carried by ships as small as 800 tons should be installed. The advantage of the VM40 is that it cuts out the need for TV tracking by using an Anglo/Dutch STIR radar which has a dual band frequency that gives very good target separation at low level.

Reaction speed

Speed of reaction is equally important. *Sheffield* was destroyed by Exocet on 4 May. Less than two and a half minutes elapsed between radar detection of an incoming aircraft and the impact of its missile. Indeed there was only a five-second interval between the first realization that *Sheffield* was under missile attack and the impact of the Exocet.

Human reactions are too slow to cope with attacks with any certainty. So Seawolf performs the whole cycle of engagement—from target acquisition to its destruction—without the intervention of human command. But, for the sake of flexibility and safety, the cycle can be over-ridden at any stage. To minimize the reaction time the Seawolf launchers have a very high slew rate. This gives high pointing accuracy so that each missile leaves the launcher in the direction of its target, saving vital seconds on the flight. In addition, the Seawolf warhead is powered to speeds of over Mach 2 by its solid booster motor

and its high explosive payload can be detonated either by an impact or proximity fuse so, although it is a remarkably accurate weapon, it can still destroy a target without hitting it.

Of all the missile systems deployed by the British fleet in the South Atlantic, only Seawolf was fully automatic. But Sea Dart was undoubtedly the most important and impressive of the earlier generation of the British sea going surface-to-air missiles. It was so impressive, in fact, the Argentine Navy had ordered some. Development work started in 1962 and the development of its command and control system—the ADAWS4—used on the Type 42 destroyers which carried Sea Dart began in 1967. The Falklands provided a battle testing for Sea Dart too. It claimed eight enemy aircraft and even dictated the course of the air battle through its one known weakness—its inability to take on very low–level targets. The basic reason for Sea Dart's success was its accuracy at great range. It has a range in excess of over 20 miles and this immediately ruled out a number of options for the Argentine Air Force. Long range surveillance planes and the high-flying Canberra bombers could not be used. They were sitting ducks. This made a part of Argentine air capacity unusable and deprived Argentine commanders of information on the whereabouts of the British fleet or its individual units. All effective attacks on British ships had to be made at very low level which brought the attacking aircraft in to danger from other anti-aircraft systems and gave them little chance of precise target identification.

Sea Dart's weakness at very low level is shared by all missiles that rely on guidance by illuminating radar. The system tracks the target by an illuminating radar and the missile homes in on the radar reflections. But the limitation of the method becomes obvious when the target is flying at wavetop height beneath the generating point of the illuminating radar signals. In this case, the radar reflections can only go downwards into the sea and it is impossible to achieve the necessary target resolution at such low levels.

Sea Dart's great accuracy above these very low levels is a tribute to its guidance system which technically makes it part of the third generation development of guided SAMs. Most of the British warships in the South Atlantic were equipped with weapons of an even earlier generation—the Sea Cat.

Human aimer

Development of Sea Cat started in the late 1950s and the first sea trials were conducted in 1962, but effectiveness does not always rely on increased sophistication and Sea Cat proved its worth by bringing down eight Argentine aircraft with two others probably destroyed. With its origin early in the guided missile era, Sea Cat relies on a human aimer. But there is no doubt that a skilled and determined operator can achieve a high degree of accuracy and the system has the advantage of relative simplicity and robustness. The early versions of Sea Cat are tracked by the operators using sighting binoculars and the missile is guided by a joystick which transmits radio commands. In a later version the operator simply has to keep the sighting binoculars on the target. Any movement of the binoculars on their mounting transmitted correction signals to the missile. After firing, a gap of about seven seconds elapses before the missile is gathered into the aimer's sight and brought under his control, which means that the system is not effective at ranges closer than 500 yards.

This does not bar it from performing as a point-defence system though. HMS *Glamorgan's* Sea Cat operator engaged the Exocet missile that struck her on 12 June.

Admittedly Sea Cat did not protect *Glamorgan* from the Exocet, but she was attacked at night in difficult aiming conditions. The fact that the system's reaction to the threat was fast enough to launch an aimed missile makes Sea Cat a worthwhile point-defence system, if not an effectively impregnable one like Seawolf.

Above Vertical-Launch Seawolf is a proposal for the future which offers even faster reaction times without the need for the launcher to track the target before the missile is released. Trials are currently in progress.

Below Three different launch configurations: top is the vertical launch unit, which allows reloads from below decks; below that is a lightweight twin-barrel launcher, again with a below-decks reload capability; bottom is the six-barrel launcher, the GWS25, fitted to 'Broadsword' class frigates.

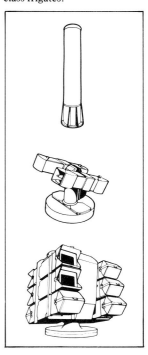

Infantry Weapons

The small arms and section weapons that both sides carried to war were more evenly matched than the Argentines care to admit. The British infantry were better equipped than their Argentine counterparts only in the matter of rocket launchers.

BRITISH INFANTRY WEAPONS

1 84mm Carl Gustav Medium Anti-armour Weapon (MAW)
2 66mm M72A1 Light Anti-armour Weapon (LAW)
3 7.62mm L7A2 General Purpose Machine Gun (GPMG)
4 7.62mm L4A2 LMG
5 9mm L2A3 Submachine Gun (SMG)
6 7.62mm L1A1 Self-Loading Rifle (SLR)
7 9mm Browning Pistol
8 7.62mm L42A1 Sniper's Rifle

ARGENTINE INFANTRY WEAPONS

1 0.5in M2 Browning Machine Gun
2 7.6mm Argentine-built FN MAG Machine Gun
3 7.62mm Argentine-built FN 50-41 Light Machine Gun
4 7.62mm Argentine-built FN 50-61 Automatic Rifle
5 7.62mm Mauser Sniper's Rifle
6 9mm Browning Automatic Pistol
7 9mm PA3-DM Submachine Gun
8 0.45in M3A1 Submachine Gun

INFANTRY WEAPONS

AMONG THE REASONS advanced by the Argentines to explain their defeat in the Falklands was the claim that the British infantry had more sophisticated weapons. They made fantastic claims for British night vision and communications equipment—claims which were not only false but ignored the fact that certain types of British equipment were inferior to their own. In fact, throughout the campaign, British infantry were better equipped than their Argentine counterparts only in the matter of anti-tank weapons.

Both sides were equipped with virtually identical versions of the most common personal weapon, the rifle. The Argentine factory at Rosario produces FN FAL NATO pattern rifles in three variations, and the Argentine infantry generally used the FN50-61, with folding stock, although some of them preferred the FN50-63, which is a little lighter and has a shorter barrel. All the Argentine versions are capable of automatic fire, but the British rifles—which were also FN FAL but in the L1A1 British type—had the self-loading mechanism but could not fire bursts. All weapons fired a standard 7.62mm round, which is accurate and effective up to 875 yards. The British had deliberately rejected the option of automatic fire when they selected and developed the L1A1. It was considered wasteful, and as the British Army is proud of a high standard of musketry training, the emphasis was placed on single-shot accuracy. The rifles and ammunition used by the army in the Falklands were not modern types, and a new re-equipment programme is in hand which will provide a new rifle with lighter rounds that 'tumble' on impact inflicting far worse wounds.

Short of ammunition

In 1982, however, the advantage may have been slightly on the side of the Argentine rifles. The Argentines had six weeks to prepare their positions and bring up adequate ammunition, enabling their automatic weapons to put down a daunting volume of fire as they reeled off bursts from their 20-round, quickly changed box magazines. The British were short of ammunition and had to march many miles into action carrying loaded weapons that were by modern standards cumbersome and heavy (10½lb). A further drawback was that the 7.62 round lacks the range, accuracy and penetrative power of old-fashioned and obsolete .303 cartridges, giving the riflemen less chance of winning a fire-fight with entrenched defenders than their fathers and great grandfathers had in both World Wars. It is also remarkable that, despite the L1A1's very low rate of fire, British soldiers were on occasions forced to use their bayonets.

Apart from the rifles, there were some lighter personal weapons distributed on both sides. For some of their officers (generally those of field rank and above), and for troops encumbered with heavy equipment, the British Army provides various types of Sterling 9mm submachine gun, which has a magazine capable of holding 32 rounds and can fire automatically or in single shots. The weapon provides great firepower but, like all submachine guns, it is inaccurate and is used principally in self defence. The Argentines have a similar tpye of weapon, the PA3-DM, which has a 25-round box magazine and a performance which is roughly equal to that of the Sterling. In addition to these standard-issue light weapons, the British special forces wielded a variety of types to suit their personal taste. Perhaps the most common was the 5.56mm Colt 'Commando' XM177, a cut down version of the M16 (Armalite) assault rifle. The XM177 has the hitting power of a rifle, but because it is a shortened version, it lacks accuracy at long ranges.

For the ordinary infantry on both sides, the first clear diversity in weapons occurred at the section level and above in the differing choice of mechanism. The MG has two functions in an infantry battalion: its first role is as a light, portable weapon adding range and firepower to each section; its second is as a much heavier backup weapon at support company level, where it can be deployed to assist the battalion with sustained fire up to considerable ranges. For many years after World War 2 the British used the Bren gun as an LMG, and various pieces of Vickers or Browning ordnance for sustained fire. By the 1960s the British had come to the conclusion that a general-purpose machine gun (GPMG) could combine the two roles of the LMG and sustained-fire (SF) support. As a result, the infantry battalions engaged in the Falklands campaign were equipped throughout—from section to support company—with a British version of the Belgian MAG, designated the L7A2 (GPMG) and firing a 7.62mm round.

The advantage of having one MG instead of two lay in the fact that every soldier could be trained to handle it, and all the ammunition was interchangeable (the Vickers Mk 1 SF weapons had indeed used .303 cartridges, in common with British LMGs and rifles, but it derived its extra range from the use of Mk 8Z ammunition as opposed to Mk 7). The disadvantage lay in the GPMG's weight—at 24lb unloaded, heavy for a section weapon—and the feeling of many was that it was not sufficiently powerful for an SF role. It is also habitually used as a belt-fed weapon, and there is a suspicion that ammunition belts snag in trees or shrubbery so that a box magazine is more suitable for an LMG. Despite these nagging doubts, the GPMG is popular with the men who use it, and it proved its value in the campaign.

First-hand experience

Experience has taught the British Army the importance of MGs. It is one of those armies—rare in the modern world—that is constantly in action, and its recent involvement in the Oman counterinsurgency campaign provided a reminder of old truths. Many of the officers who went to the Falklands had first hand experience of war, and they made sure they took with them every MG on which they could lay their hands. This meant that there were dozens of GPMGs taken into each battalion attack in their bipod-supported LMG role, and a fair number with each support company mounted on superb buffered tripods to exploit their range of 1300 yds and high rate of fire of 750-1000 rounds per minute in the SF role.

In addition to GPMGs there were older MG types ransacked from stores by resourceful British soldiers. There were a number of Bren LMGs, that 'perfect weapon of war', used as personal weapons despite their individual weight of 21lb without a 30-round box magazine. To add weight to the SF

Top A Royal Marine fire trench overlooking San Carlos. The marine on the left crouches over the sights of his L7A2 GPMG. He has taped a 'Snoopy' doll to the carrying handle! For British soldiers, their personal weapon is the most important thing in the world—it receives even more care than the soldier himself. Contrast this with the captured Argentine weapons (**above**), some of which are not very clean. Both sides used very similar small arms—the major difference was in the quality of the men with their fingers on the triggers.
Right A paratrooper cradles his SLR as he enters Port Stanley.

Of course, the Argentines had less cause to use anti-tank weapons as the British maintained manoeuvrability over open ground, but they did site their 105mm Recoilless Gun Model 1968 to great effect—particularly in the battle for Mount Longdon. However, they must have been stunned by the liberality with which the British used anti-tank rounds against their sangars and entrenched strongpoints. The British weapons were the Swedish 'Carl Gustav' 84mm recoilless rifle which fires 6½lb rounds, and the US made 66mm LAW one-shot throway assembly package, each one of which weighs only 5lb. The 84mm 'Charlie G' is very accurate. With an X3 magnification telescope sight and an effective range up to 765 yards, its projectiles can burn a hole through 1½in of armour plate, which gives them plenty of power to break up bunkers. The 66mm LAW is obsolescent because it is not capable of taking on the Main Battle Tanks of the Warsaw Pact, but it proved good enough to chew through Argentine infantry strongpoints and it is accurate up to 165yds

Ferocious determination

Some observers drew the conclusion that rocket launchers gave the British infantry their advantage over the Argentines. The credit, however, lies equally with the ferocious determination of the British infantrymen, who carried such a great weight of firepower in machine guns and anti-tank weapons into action, and then had the skill to use it properly.

Even this firepower did not silence all the Argentine positions, and some more old-fashioned weaponry had to be used in close quarters combat. There is probably not much to choose between the British L2 grenades—nowadays invariably thrown rather than rifle launched—and the Argentine CME-FMK2-MO. But grenades are most effectively used to winkle enemy out of entrenched or enclosed positions and, during the Falklands campaign, it was the Argentines who occupied the trenches and the British who were winkling them out. As a result, the most effective use of grenades was made by the British. In the final count the British relied on bayonets, and it is something of a curiosity that the British are still so wedded to the idea of using this ancient but deadly weapon.

Indeed, it is in the philosophy behind their use of infantry weapons that the British proved themselves so superior to the Argentines. They believed, correctly, in winning the fire-fight with a heavy volume of fire. At section level the weight of fire from their machine guns and rocket launchers proved decisive. It is worth mentioning, however, that although the Argentines seemed particularly vulnerable to night attacks, this was not because of the technical superiority of British night vision equipment. Both sides used image intensifiers which gather in whatever light is illuminating the target (starlight or moon light) for no night is actually pitch dark, and amplifies it to contrast the target with its surroundings. The British devices used by the infantry were so-called 'first generation' equipment capable of providing their snipers and observers with much better vision.

A contrast between the equipment of the Argentine and British infantry equipment does not show any great British advantage. If there was a difference, it lay in the men using it.

forces some Browning heavy-barrel M2s were pressed back into service with their 50-calibre ball ammunition and stolid 450 to 500 rounds per minute rate of fire. However, the Browning weighs 84lb even without a suitable tripod mounting, and it was used principally to add to anti-aircraft defences. The inclusion of so much extra automatic fire from genuine machine guns proves that the British sensibly attached great importance to overwhelming their adversary in a fire-fight.

In contrast, the Argentines stuck to the principle of using LMGs at section level and backing them up with SF from heavier calibre support weapons. The LMG is simply their rifle with a heavy barrel and attached bipod, designated the FN50-41. Purists would probably insist that the FN50-41 is a machine rifle rather than a genuine machine gun, because although the heavy barrel means that the weapon can maintain a higher rate of fire than the automatic rifle, there is no improvement in range nor much in accuracy. While the British GPMG is by far the most important source for the generation of firepower in a section, the Argentine FN50-41 gunner could not provide such comforting stopping power for his immediate comrades. The Argentine weapon has the attraction of being very light at 13lb unloaded, but would come second best in a performance contest with the GPMG.

With such a light section weapon, the Argentines were forced to provide SF backup with much heavier MGs—.5in and some .3in Browning MGs. The .3in is obsolescent but it is still an excellent weapon, and both calibres are alike in construction and performance.

The Argentines were not glaringly deficient in MG support, although they experienced some problems caused by the use of different ammunition types. It was in the provision of rocket-launched anti-tank projection that their ground forces were completely outmatched by the British. Accurate medium-range anti-tank weapons are distributed among Argentine infantry only at a support level, while British infantry bristle with powerful rocket launchers in every section.

On track for war

One of the major equipment successes of the Falklands campaign was the Combat Vehicle Reconnaissance (Tracked)—better known as the Scorpion and Scimitar, fast, flexible gun platforms.

Top During the advance on Douglas, Scorpions and Scimitars like this one were used to speed progress by acting as a taxi service. Normally these CVR(T)s only carry a crew of three—a commander who also loads the guns, a gunner/radio operator and a driver. The first two sit side by side in the turret while the driver sits below in the hull.
Above A combat-loaded Scimitar is driven onto a landing craft during the crossdecking operation at Ascension. Originally, the CVR(T) was designed to be light enough to be dropped by parachute.

THE ORIGINS OF THE CVR(T) lie in a British Army requirement drawn up in the late 1950s for what was called an Armoured Vehicle Reconnaissance. The AVR should combine the roles of reconnaissance and fire support, as well as having anti-tank capability and being light enough to drop by parachute. This limited the all-up weight to 15,000lb and made it impossible to design one vehicle to cover all these roles. So the designers came up with two versions—a tracked model with 76mm gun, which became the CVR(T) Scorpion, and a wheeled vehicle carrying a 30mm cannon the CVR(W) Fox.

While the Scorpion was being developed and produced, other variants were being designed. First off the line was the Scimitar, which is very similar to the Scorpion but carries the 30mm Rarden gun. Then came the Striker, which carries the Swingfire anti-tank guided missile, and the Spartan, which was basically an Armoured Personnel Carrier but used in a number of roles in the British Army. Three other types were also produced: the Samaritan, an armoured ambulance; the Samson, a recovery vehicle; and the Sultan, an armoured command vehicle. The types used in the Falklands were the Scorpion and the Scimitar, with one Samson providing recovery back up.

As a reconnaissance vehicle the CVR(T) should have maximum mobility—not just speed but also the ability to traverse all types of ground. Armoured fighting vehicles often have trouble with soft going. To overcome this, it is necessary to keep what is called the nominal ground pressure—that is the vehicle weight divided by the area in contact with the

ground—to a minimum. In other words, it is a question of limiting the weight and spreading it as widely as possible. This has been done by using aluminium armour which gives a vehicle weight of just over 17,500lbs and putting tracks on the CVR(T). But there was still much doubt whether it could cope with the Falklands boggy terrain.

'Cowpats' of explosive

Although the infantry had Milan anti-tank guided missiles and smaller infantry anti-tank weapons, these did not have the flexibility of a mobile anti-tank gun platform. Here the Scorpion and Scimitar had something to offer. Scorpion's 76mm gun fires an anti-armour round, the High Explosive Squash Head. On landing, this forms a 'cowpat' of explosive which then detonates. The shock waves go through the armour and dislodge pieces from the insides. These fly around causing much damage to the crew. The HESH round is effective against all but the heaviest tanks to a range of 1500yds, and it can be used in the ordinary High Explosive mode. The CVR(T) also carries straight HE rounds, a canister which operates just like a shotgun cartridge for use against infantry at close quarters, illuminating and smoke rounds.

The 30mm Rarden gun on Scimitar can fire a single shot or three-round bursts. Its Armoured Piercing Special Effects round penetrates armour, then explodes inside and is designed for use against APCs. It also has a straight armour piercing and an HE round. Between them, these two guns can take on a wide variety of targets.

Below The Blues and Royals soon found that their Scorpions were not confined to established tracks but could roam anywhere.
Bottom Scorpion 76mm gun can fire a High Explosive Squash Head round, which is effective against all but the heaviest tanks to a range of 1500yds. Built by Alvis of Coventry, all the variants of the CVR are based on the standard Scorpion chassis and make maximum use of interchangeable parts.

The decision to take over CVR(T) to the Falklands was made on 3 April. But lack of space on board ship and doubts about their capacity to cope with the conditions meant that a limited number were taken. Only four Scorpions, four Scimitars and one Samson were embarked. They were organized into two troops, with the standard mix of two Scorpions and two Scimitars in each. The normal tactic is to operate in pairs, a Scorpion and Scimitar together.

At San Carlos Bay, there was still a certain amount of confusion about how the vehicles could be used. But, while plans for the advance were being consolidated, the CVR(T)s found themselves in an unexpected role—ferrying stores forward. There was a lack of mechanical transport and vast quantities of

stores being landed. These had to be got forward, and the CVR(T)s provided a means of doing this. In view of their shape, they could not carry much, but every little helped.

With the breakout from the beachhead, the CVR(T)'s supported 45 Commando and 3 Para in their advances towards Douglas and Teal inlet. Surprisingly, the Blues and Royals found that they could roam almost at will across the peaty ground. There was one occasion when a vehicle commander jumped from his CVR(T) and sunk up to his knees in a bog, yet the vehicle itself had not even broken the crust. This meant that the vehicles could be used in their true role in the advance, going on ahead of the marines and paras, who were on foot, and getting up on high ground to cover the advance forward by securing the flanks.

Estancia to San Carlos

After the taking of Estancia House, Major General Jeremy Moore began to realize just how useful the CVR(T) could be and he decided that both troops should join up with 2 Para at Bluff Cove. Headquarters estimated that it would take 36 hours for the troops to make their way across from Estancia backtracking via San Carlos along proven routes. But by then the Blues and Royals had built up a great deal of confidence in their vehicles and elected to take the supposedly impassable track south from Estancia House. They made the journey in just six hours.

One CVR(T) troop was put under command of the Scots Guards for their attack on Tumbledown. Part of the plan was to put in a diversionary attack along the Bluff Cove-Stanley route to cover the move of the battalion from Goat Ridge westwards to the foot of Tumbledown. It was decided that the Blues and Royals troops should lead the diversionary attack to draw the enemy fire.

The operation began after dark on 13 June but the leading Scorpion ran over an anti-personnel mine and was put out of action. There were no injuries to the crew. The remainder of the troop then stayed where it was and provided fire support.

Meanwhile, the other CVR(T) troop went with 2 Para to Wireless Ridge. Here again the Blues and Royals were used very much in the fire support role. By now, the troops had evolved the technique of using their 76mms and Rardens to engage bunkers, dug-outs and outcrops of rock and flush out the enemy who would then be taken on with the coaxially mounted general purpose mounted machine gun. The mobility of the vehicles meant that they could deploy quickly to a spot that was giving trouble and their night vision devices meant that the darkness was no hindrance.

The CVR(T) also had made a contribution to the final battles of the campaign and by the end of the fighting, each vehicle had covered almost 400 miles over very difficult going. The crews themselves kept up maintenance and there was only one major assembly failure, a gear box.

The Scorpions had fired an average of 60 rounds of 76mm each and 120 rounds had also been fired from each Scimitar's Rarden. With the benefit of hindsight, there is no doubt that more CVR(T)s would have helped the campaign. As it was, three members of the Blues and Royals were mentioned in dispatches and appeared in the Falkland Islands honours list.

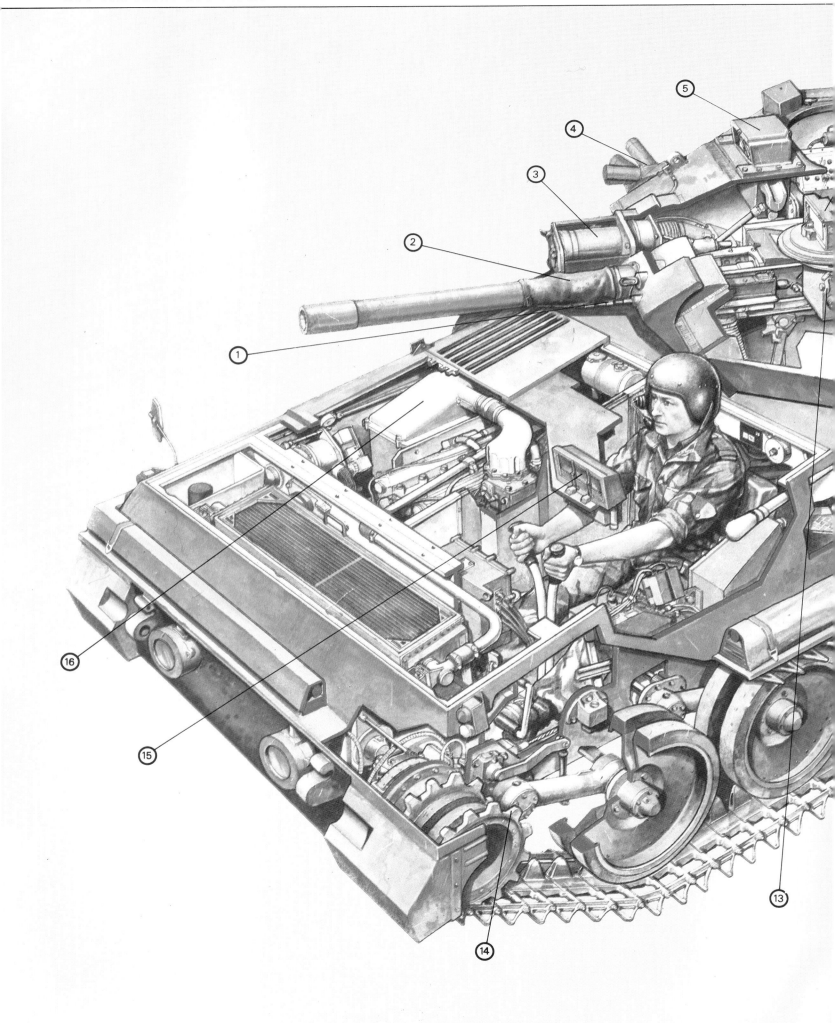

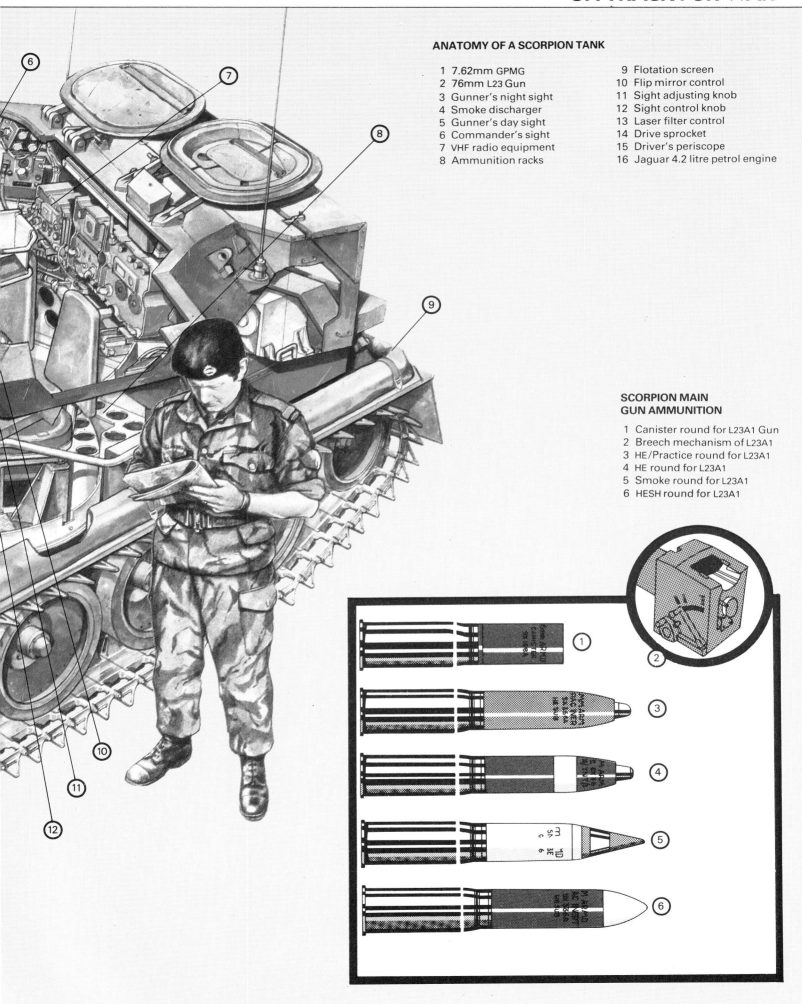

ANATOMY OF A SCORPION TANK

1 7.62mm GPMG
2 76mm L23 Gun
3 Gunner's night sight
4 Smoke discharger
5 Gunner's day sight
6 Commander's sight
7 VHF radio equipment
8 Ammunition racks
9 Flotation screen
10 Flip mirror control
11 Sight adjusting knob
12 Sight control knob
13 Laser filter control
14 Drive sprocket
15 Driver's periscope
16 Jaguar 4.2 litre petrol engine

SCORPION MAIN GUN AMMUNITION

1 Canister round for L23A1 Gun
2 Breech mechanism of L23A1
3 HE/Practice round for L23A1
4 HE round for L23A1
5 Smoke round for L23A1
6 HESH round for L23A1

Conqueror and Fearless

While the nuclear submarine *Conqueror* kept the Argentine Navy bottled up in port, the assault ship *Fearless* was saved from the scrapheap to play a vital role in the landings at San Carlos. Both ships were essential to British victory in the Falklands.

Below A rare shot of *Conqueror*. The modern nuclear submarine is designed to stay under the water, unlike conventional diesel-powered O-class boats which spend much of their time on the surface.

ALTHOUGH THERE WAS ONE ENCOUNTER between a British nuclear submarine and the Argentine Navy—*Conqueror*'s sinking of *General Belgrano*—their presence, or suspected presence, was enough to reduce the sizeable and comparatively well-equipped Argentine fleet to total impotence.

Though the fear they instil in conventional navies is massive, nuclear submarines themselves are, surprisingly, not large. With an overall length of 285ft, *Conqueror* is over 10ft shorter than the conventional O-class boats. She is, however, rather wider—33.3ft compared with 26.5ft. Nuclear submarines are designed to give optimum submerged efficiency, their virtually unlimited endurance enables them to remain submerged for the complete duration of a patrol, while the design of conventional submarines dates from the time when the air-breathing boat needed to spend a significant proportion of its time on the surface.

The more portly proportions of the nuke enable her to have the maximum internal volume for a minimum surface area wetted. *Conqueror*'s dimensions give her a submerged displacement of 4900 tons, over twice that of the longer conventional O board, and she carries a crew of 103, as against 69.

CONQUEROR

Laid down in 1967 and completed in 1971, *Conqueror* belongs to the trio of C type follow-ons to the Valiants, themselves largely based on the prototype *Dreadnought*. Few details of the reactor have been released, but it is of the pressurized-water type. Heat generated by the nuclear reaction is carried by continuously circulating water through a heat exchanger where the secondary output is used to raise steam in a conventional boiler. This drives a turbine which is coupled to the single main shaft through reduction gearing. The actual power at the propeller is given as 15,000 horsepower. An electric motor and battery are also fitted for emergency use.

The use of nuclear power allows non-stop high speed submerged passages without the need for re-fuelling. This enabled *Conqueror* and her companion boats to arrive in the disputed zone and establish a *cordon sanitaire* well before the arrival of the Task Force. High speeds must be used circumspectly, though. They generate noise and the world of submarine warfare is one of silence. High submerged speeds also demand fine control. The boat has to be 'flown', rather like an aircraft, in a comparatively narrow channel between the surface and the safe working depth. At maximum speed of 28 knots, the boat is travelling at over 47ft/sec and to be safe, she needs plenty of searoom. One of the great limitations of the Falklands operations was that they were conducted in the shallow water of the continental shelf. *Conqueror* carried a variety of weapons. Through her six-bowed mounted tubes, she could launch an encapsulated Harpoon anti-ship missile, a long-range Tigerfish torpedo or a short-ranged Mark 8 torpedo.

In the submarine-launched form, Harpoon is fired from the tube in a capsule which is blasted to the surface. This then falls away leaving the missile to follow its pre-programmed track, finally using active homing to close on its target. It has a range of up to 60 miles, nearly twice an Exocet's. It also follows a wave-hugging trajectory and delivers a 500lb warhead, compared to Exocet's 363lb.

A Tigerfish torpedo belongs to the current generation of large weapons with a good endurance and control. Though not officially announced, its range is at least 20 miles, and it can be launched on a pre-programmed search pattern to acquire the target, when an active homing head takes over. A guide wire paid out behind the torpedo allows for a two-way exchange of data between weapon and submarine so if it is attracted by the wrong ship or seduced by a decoy, its guidance system can be over-ridden.

But it was the Mark 8s that were actually used to sink *Belgrano*. These were old and tried weapons but lacked range and an approach to within three miles would probably have been necessary. The escorting destroyers were of World War 2 vintage—ex US naval ships with neither stand-off anti-submarine weapons, such as ASROC, or a helicopter. The risk was acceptable as long as a high-speed disengagement could be made.

Although nuclear submarines were used to escort the Task Force, their major function is to prey on their own kind, contesting control of the North Atlantic with Warsaw Pact fleets.

A problem facing NATO's nuclear submarines is that, with missiles of ever-increasing range, the Soviet ballistic missile submarines need hardly leave their own doorstep. Missile carrying submarines can lurk under the polar-ice cap surfacing in pools or thin patches to launch their weapons. They can hang there silently and almost indefinitely. Any submarine seeking them needs to tread carefully. Contact with the ice roof—which may be flat and only a few feet below the surface or hung with

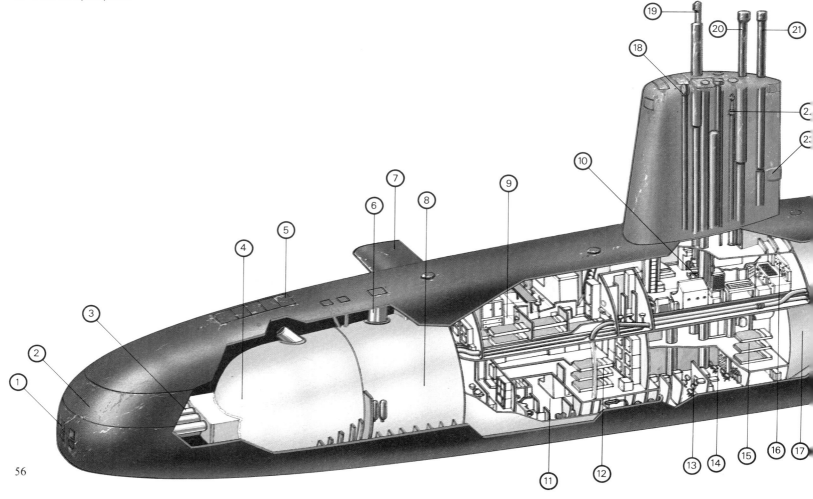

TIGERFISH

1 During the launch phase, a command guidance wire is unreeled from both the torpedo and the submarine.

2 Guidance commands are generated by the submarine's computer system and are passed to the torpedo's control surfaces by its own computer.

3 When the torpedo is within range of its target it passes into the terminal phase. Guidance control passes to the weapon's onboard sonar system. But at all times the submarine can override the torpedo's computer to abort the attack.

4 Onboard sonar guidance system.

5 Pop-out wings provide roll stabilization.

6 Counter-rotation propellors.

7 Directional control surfaces.

8 Onboard wire reel.

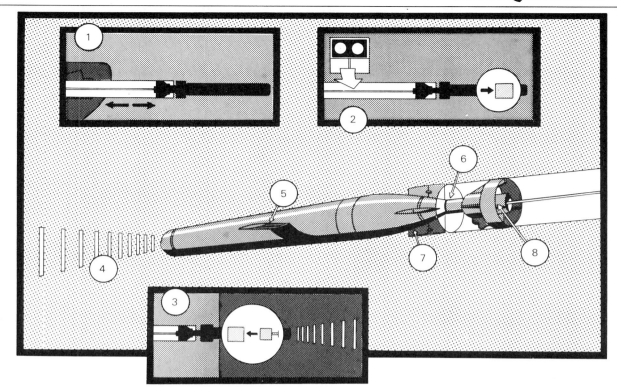

pinnacles, even reach the bottom in pressure ridge conditions in shallower water—could be fatal. High definition sonars can map the underside of the ice-cap but advertise the hunter's presence to a listening, motionless quarry.

But nuclear submarines' main drawback is their cost, which inevitably limits the numbers of complementary surface warships that can be built. Though in some respects less capable, a surface warship is far more flexible than a submarine as it can grade its response to a particular threat. A nuclear submarine can disguise its presence or it will try to kill—it has no other option.

Fearless

More functional than elegant, *Fearless* is best visualized as a combination of floating dock and ship. Together with her younger sister, *Intrepid*, she

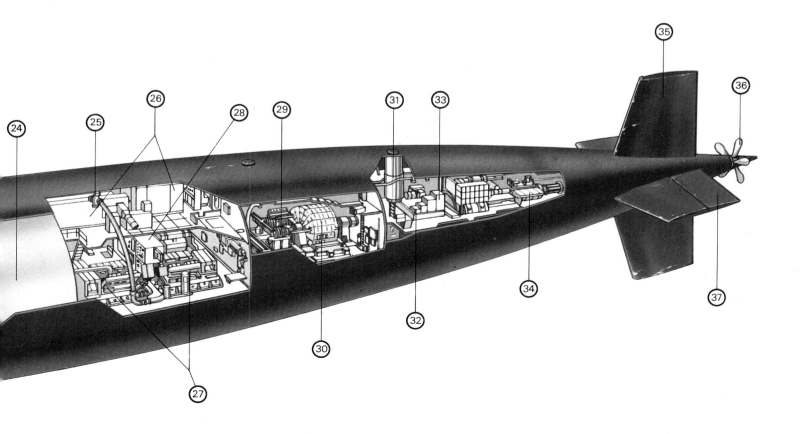

FEARLESS

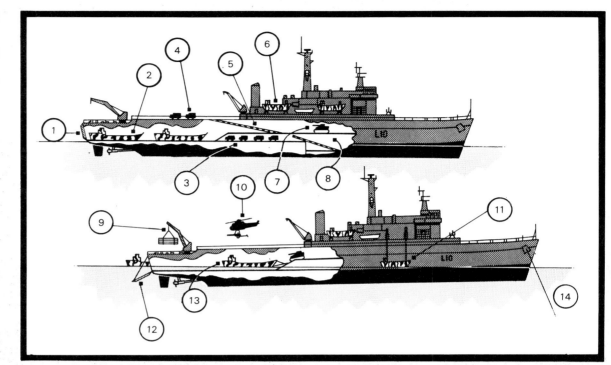

FEARLESS IN OPERATION

Seagoing Condition

1 Stern gate raised to seal entrance to dock.
2 Four LCM9s, two by two, pre-loaded with two MBTs or equivalent.
3 Half deck carrying 20 LWB Land Rovers or equivalent.
4 Additional wheeled transport on upper deck.
5 Lifting ramps interconnecting vehicle decks.
6 Four LCVPs in davits.
7 Tank deck which can carry up to 15 MBTs.
8 Additional storage.

Docked Down for Amphibious Operation

9 Cranes off-load heavy gear into LCVPs or LCMs alongside.
10 Helicopters carry troops or equipment directly ashore.
11 LCVPs lowered from davits.
12 Stern gate lowered.
13 LCM9 loading MBT from tank deck via internal beach.
14 Ship anchored. Two stern anchors also available for use in difficult conditions.

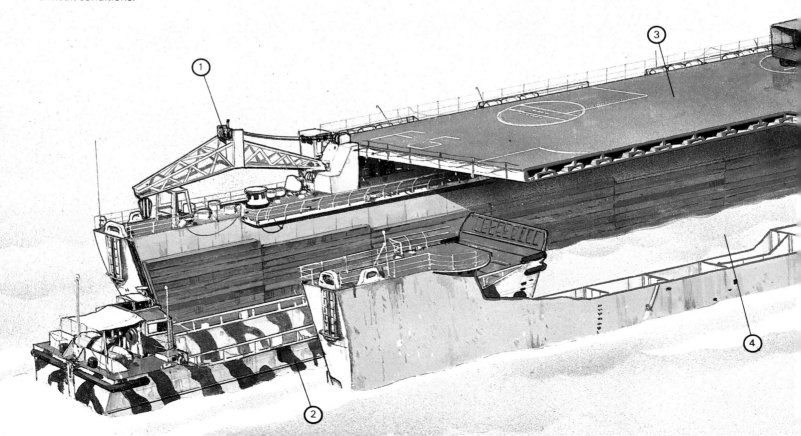

ANATOMY OF HMS *FEARLESS*

1 Dock crane
2 Landing craft (LCM9) entering dock
3 Helicopter deck and vehicle park
4 Flooded dock
5 Sea Cat missiles
6 Vehicle stowage decks

7 Offset funnels
8 Landing craft (LCVP)
9 Radar and radio equipment
10 Command HQ and operations room
11 40mm Bofors

Below In normal seagoing conditions the level of the deck in *Fearless*'s docking well is above the waterline and acts as a freeboard deck which is self-draining if the stern gate is broached. But here the stern gate has been lowered and the docking well flooded, giving the ship a stern heavy trim. In seagoing trim, the bow draught is 23ft. When the docking well is flooded, the bow only draws 20.5ft. While the landing craft ferry troops and vehicles from the flooded docking well, fleets of helicopters lift other equipment ashore from the flight deck aft.

is often termed an assault ship, but this is a loose term covering many types, and she can be more accurately described as an amphibious transport dock, Landing Platform Dock.

The origins of *Fearless* lie in the tank landing craft of World War 2, little more than robust shoe boxes with large box doors which gave the Allies the ability in Europe and the Pacific to put troops and equipment over the beach rather than rely on the acquisition of a major port. The Landing Ship Dock represented a further development, in effect a self-propelled floating dock with rugged seakeeping ability which met the requirements of long-range operations. In place of a dock's open ends a bow was added forward and a large bottom-hinged gate aft. Superstructure bridged the docking well at the forward end and machinery was added in the narrow side spaces flanking the dock.

Though looking very similar externally, the LPD incorporates a shorter docking well. The length of that in *Fearless* has not been released, but, in order to house the four LCMs in a two-by-two stowage, it must be at least 180ft × 45ft. The smaller dock makes

more space available for the accommodation of troops, their armour, wheeled vehicles, ordnance and stores.

The layout of the ship is complex and has much in common with a modern roll-on, roll-off ferry. Of the main hull, only the section from the bridge forward is devoted to the needs of the ship's 570 personnel. That part of the ship aft of the starboard (after) funnel approximates to the dock, with vehicle and stores stowage in the centre third. Troop accommodation is largely in the lower superstructures.

The ship's main landing craft are four LCM9s. These Landing Craft, Mechanized are 85ft vessels with a working deadweight of about 100 tons, which can comprise wheeled vehicles or two main battle tanks.

To undock the LCMs, *Fearless* takes aboard nearly 5000 tons of water, giving her a heavy-stern trim. At this stage, the deck of the docking well (essentially the freeboard deck and, in normal trim, above the waterline) is well submerged by water that has been admitted. Once trimmed down, and with equal water level inside and out, *Fearless* can lower her

sterngate and allow the craft to back out. Once they have left their loads ashore, the craft return for more.

They can be admitted to the flooded well two abreast and lay their bow ramps against a false steel beach that leads directly from the tank deck enabling armour to board the smaller craft directly. Other wheeled vehicles and artillery can transfer to this deck via ramps from their stowage on other levels, including the open deck spanning the after end of the well.

Beneath heavy davits, two to a side, *Fearless* carries four LCVPs (Landing craft, Vehicle Personnel). The Mk2 versions that are shipped have a 5-ton capacity that can be used alternatively for either 35 fully-equipped troops or two Land Rover-type vehicles. Cranes on the quarters can transfer stores directly to craft alongside.

A further aid to the rapid establishment of a bridgehead is *Fearless'* ability to spot up to five Wessex-sized helicopters on the after deck. No hangar space is provided as such a flight is attached only temporarily for particular missions, but troops and light support landed by these choppers would have the task of establishing temporary command over a limited area that would then be rapidly reinforced over the beach.

The extremely tall mainmast supports the necessary communications aerials which are a vital part of *Fearless'* role as a headquarters ship. Her bulky superstructure houses the personnel and equipment to direct an amphibious landing and to enable an army brigade headquarters to control operations temporarily until it can go ashore. *Fearless* carries a quadruple Sea Cat launcher on each corner of the superstructure, single 40mm guns in the bridge wings, two nine-barrelled launchers and infra-red flares.

At the time of the South Atlantic emergency, both *Fearless* and *Intrepid* were slated for early disposal, though only 15 years of age. Their obvious indispensability in a balanced fleet has earned them a qualified reprieve. With both the US Marine Corps and Russian Naval Infantry using several landing ships between them, however, the role of the landing ships has never been stronger. It seems inconceivable that *Fearless* and *Intrepid* should be consigned to the breaker's yard.

Sidewinder, Shrike and Paveway

For the first time the RAF and Fleet Air Arm used Sidewinder, Shrike and laser-guided bombs in anger. We look at how they worked, and how these weapons evolved. From Vietnam to the South Atlantic, they have achieved an enviable record.

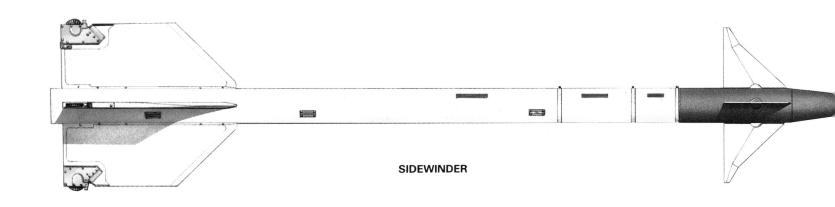

SIDEWINDER

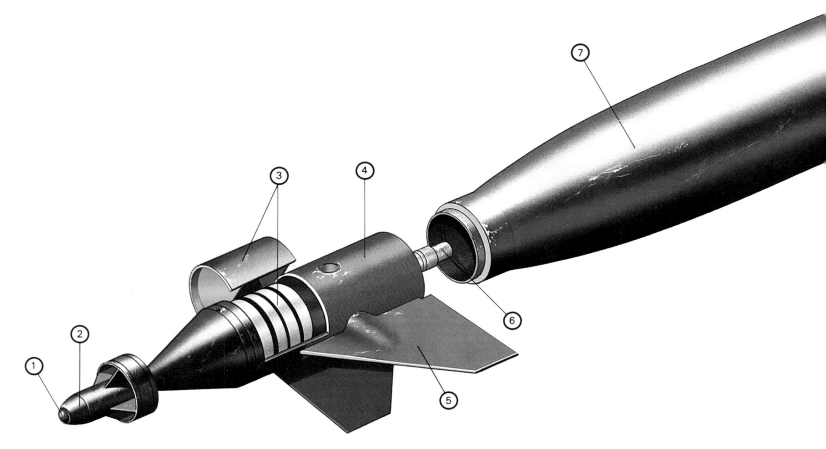

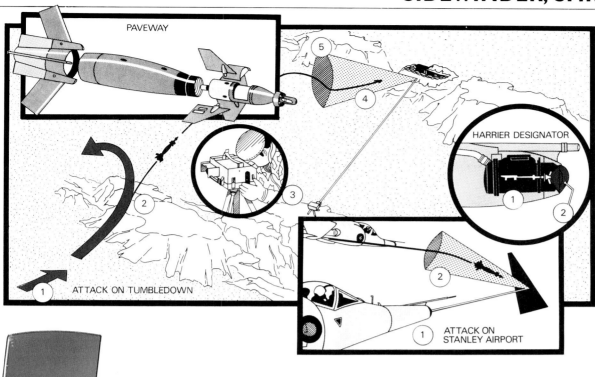

PAVEWAY

HARRIER DESIGNATOR

ATTACK ON STANLEY AIRPORT

ATTACK ON TUMBLEDOWN

ATTACK ON TUMBLEDOWN

1 Harrier approaches the target low and fast from the SW
2 Harrier pulls up into a 30°climb and launches a paveway upwards
3 Ferranti laser ground designator 'illuminates' the target
4 Having climbed to 1500ft, the Paveway falls back into the laser 'basket' and homes in on the target
5 Laser 'basket' reflected by the target

HARRIER DESIGNATOR

1 Seeker and laser firing head
2 Protective 'eyelid' shutter

ATTACK ON STANLEY AIRPORT

1 Designator Harrier illuminates Stanley's runway
2 Paveway Harrier launches a weapon into the laser 'basket'

PAVEWAY

1 Optics/silicon detector
2 Aero stabilizing detector
3 Computer and housing
4 Control section, hot gas generator, thermal battery
5 Control fins (4)
6 Fuse
7 Warhead
8 Wing release mechanism
9 Wings extended (4)

SHRIKE

SIDEWINDER, SHRIKE, PAVEWAY

Below The largest fighter in the world—an RAF Nimrod with four Sidewinders under its wings.
Right Sidewinder in action: the sequence shows a missile fired from a Sea Harrier shooting down a Skyhawk.

THE AMERICAN AIM9 SIDEWINDER is the father of a whole generation of air to air guided weapons.

The AIM9 family comprises 13 different models, three of which, the AIM9B, 9G and 9L were used by the British forces in the Falklands. The first of these was the earliest production model and was made for the European NATO countries by a consortium headed by the German firm of Bodenseewerk Gerätetechnik. In its standard form, AIM9B is 9ft 4in long, 5in in diameter with a span of 1ft 10in.

Directional control is provided by two sets of fins mounted at the tail and just aft of the nose, and the weapon is powered by a Rocketdyne solid fuel rocket motor which produces a thrust of 600lb. Guidance is provided by an infra-red homing system which locks onto the heat of a target's engine exhaust. AIM9B's destructive power is provided by a 24lb high explosive charge. It has a maximum speed of Mach 2.

If the Falklands War produced any secret weapon, then the Royal Navy's AIM9Ls must be a strong contender for the title. Identifiable by its double-delta forward control fins, the AIM9L is a much more effective weapon that the earlier models used by the RAF. The secret of its improved performance lies in its use of the AN/DSW29 guidance and control system which is credited with 'significantly enlarging the firing envelope'.

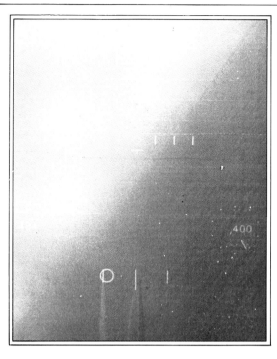
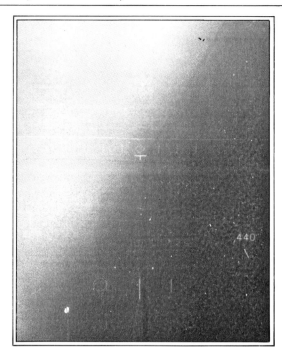

Far left The AGM45 Shrike, which was used to shoot up Argentine radar installations. It met with mixed success, but will certainly be used again. **Above** The brains of the laser-guided Paveway bomb: the laser seeker attached to its nose. The bomb homes in on the laser light reflected from the target and follows it all the way to the point of impact.

The 'firing envelope' is the area in which the weapon's infra-red homer is effective. In the earlier model AIM9s, this was a restricted cone with the target's engine exhaust at its centre point, and meant that the interceptor had to be almost directly behind the target to ensure the missile's homer locked onto it. Combat operations, especially in Vietnam, proved that a target which manoeuvred violently could break the lock quite easily.

In the AIM9L, the sensitivity of the guidance system is such that a launch platform does not have to be so rigidly positioned behind a target to ensure a hit. And the redesigned control surfaces have made it much more manoeuvrable.

The use of the AIM9L in the Falklands came as something of a surprise. All such weapons held in the UK were part of NATO war stocks, that is, arms which are supplied solely for use in NATO operations. Speedy and secret re-supply from the US enabled these stocks to be depleted during the campaign.

So unbeknown to anyone—least of all the Argentines—the Task Force's Sea Harriers went into action, armed with the very latest Sidewinder. Throughout the campaign, 26 AIM9Ls were fired in anger, of which 25 functioned perfectly. Some 16 resulted in confirmed 'kills' taking out 11 Mirages, 4 Skyhawks and a Canberra, and one probable 'kill'.

The first Shrike was developed by the US Navy during the early 1960s and is designed to knock out enemy ground radars. The AGM45 is a 10ft long missile with four stabilisers grouped around the tail and four control surfaces, giving it a span of 3ft mounted at the mid-fuselage point.

The body of the weapon has a diameter of 8in and the whole system weighs 390lb. The high explosive fragmentation warhead alone weighs 145lb. The Shrike is powered by a Mk39 Mod 7 or Mk53 solid fuel rocket motor which gives the weapon a maximum speed in the region of Mach2 over a range of 7 to 10 miles. The heart of the Shrike is a receiver unit which is tuned to the frequency of the radar under attack.

Combat use showed the weapon to have a number of shortcomings. Its guidance system can be knocked out by simply switching off the target radar for a short period, for example, leaving the missile with no signal to home on to. The same effect is also created by the use of 'frequency agile' equipment.

The AGM45 is not normally in the RAF's weapons inventory and was used as a last minute replacement when it was suspected that Anglo-French AS37 Martel was inappropriate to the conditions.

The Texas Instruments' Paveway is not a missile but a guided bomb. It comprises the explosive section of a standard free-falling bomb mated to a tail section carrying four pop-up wings and a nose unit which carries four control fins, a computer and a lasar-seeker head.

Operationally, the Paveway system comprises two distinct units, the weapon itself and a separate laser designator. The designator may be carried in another aircraft or on the ground. In either case, the unit is used to direct a beam of laser light at the target. The Paveway, once free of its carrier, searches for the laser light reflected from the target with its seeker head. The weapon locks on to the light source and the nose-mounted computer steers it with the control fins towards the target.

During the Falklands campaign, Paveway was used operationally by No 1 Squadron who attempted to use Paveway bombs against Port Stanley airfield early in the fighting before the designators arrived. One Harrier would dive at the target while the second Harrier illuminated the target with its on-board laser range finder.

By the beginning of June 1982, the designators finally caught up with the forward troops and were used for the first time on the 13th. During the morning two aircraft attacked Argentine positions on Mount Tumbledown. The first Paveway fell short as the designator was used too soon. But a second made a direct hit.

No 1 Squadron was back in action during the afternoon aiming this time at a heavy machine gun nest or a 105mm howitzer emplacement on Tumbledown. Again the first Paveway fell short and the second made a direct hit.

Regiments and Regalia

Badges, insignia, and uniforms — the very symbols of military tradition. Like most organizations of long historical standing, Britain's Armed Forces are made up of a number of highly individualistic elements, each one reinforcing the sense of unity and discipline inherent in Service life.

Each Regiment and branch of those Services contributed its own unique brand of courage and determination in achieving the final victory.

Operation Corporate, the combined service operations that liberated the islands, rested on the shoulders of some of the best-trained fighting men in the world. In this revealing series of profiles, we present the background story of those who went with the Task Force, and the proud units they represent.

ROYAL MARINES

No fewer than 36 countries in the world maintain marine corps or forces of naval infantry, ranging in size from Guatemala's 200 marines to the mighty 188,900-strong US Marine Corps. But none is older or more experienced than Britain's Royal Marines—7899 men in all when the Falklands crisis called on them for a unique display of their training and highly specialized skills.

Descending from the 'Soldiers at Sea' of 1664, the title 'Royal' was awarded in 1802 and the Commando (Cdo) role came in World War 2. The Royal Marines took responsibility for the role in 1946. Since 1974, 3 Cdo Brigade (Bde) had been committed to NATO's Northern Flank (Norway and Denmark), arctic warfare training having begun in 1970.

The Cdo bde consists of 40, 42 and 45 Cdos, each the equivalent of an Army infantry battalion with 650 men organized in 3 rifle coys (each of 3 troops), a support coy (MG, mortars, assault engineers, snipers and Milan anti-tank missiles) and an HQ. Number 41 Cdo was disbanded in 1981. In support are 29 Cdo Regt, RA (18 105mm guns); the Cdo Logistic Regt (formed in 1972); an air sqdn (18 helicopters); air defence troop (12 Blowpipe missile launchers); 1 Raiding Sqdn (17 Rigid Raider and 16 Gemini craft); the Special Boat Squadron (SBS); 59 In-

This marine is kitted out with the sort of gear that typified the earlier stages of the Falklands war, before troops started to wear the 'Arctic' kit and became loaded down with enormous packs. He is wearing: dark green marine commando beret with blackened cap badge; windproof anorak over DPM jacket; cold-weather mittens with separate trigger-finger; '58 webbing with respirator in its case on left hip (hanging from belt); field dressing attached to shoulder harness; bandolier of extra ammunition across chest. His weapon is the 9mm Sterling sub-machine-gun (SMG), which would indicate that he is a senior NCO, although no rank badges are worn on windproofs.

dependent Cdo Sqdn, Royal Engineers; a Mountain and Arctic Warfare Cadre and Cdo Forces Band (medical duties).

Equipment is similar to the Army's and on the same scale, but there are special items such as the tracked Volvo BV202 snow vehicle nicknamed 'Bandwagon' and used in the Falklands.

Every RM recruit undergoes a 32-week course at the Commando Training Centre, Lympstone, Devon. Only on passing does he bear the unit's green beret. Officers have a year's training including Lympstone (they have to complete the 30-mile speed march in 7 hours to the recruits' 8). Lympstone also runs the NCO promotion courses, the PT School and drill training. Heavier weapons and specialist training is done with the Army or in the case of helicopter pilots with the RAF. The RM School of Music at Deal, Kent, turns out buglers (bugles and drums) and musicians (all other instruments).

Amphibious Training Centre, Poole, Dorset, trains landing craft crews (volunteers from corporal upwards), detachments (usually about 10 men) for service in frigates, and craftsmen of all types. SBS entry is for already qualified marines and includes a week's diving in the aptitude and fitness test. A 15-week training course takes in seamanship, boatwork, diving and demolition followed by a 4-week parachute course. The successful marine becomes a Swimmer Canoeist, Grade 3. Unlike his SAS counterpart he remains permanently with the unit. The main training for 3 Cdo Bde comes in Jan-March for mountain and arctic warfare in Norway. Every marine is a qualified skier and trains alongside 1 Amphibious Combat Gp of the Royal Netherlands Marine Corps.

The Royal Marines won 46 decorations, including 2 DSOs, a DFC, a DSC, 6 MCs and 11 MMs, and 83 mentions in dispatches for the South Atlantic Campaign. Casualties totalled 27 killed and wounded; 45 Cdo lost 12 killed.

PARACHUTE REGIMENT

The Parachute Regiment only dates from 1940, but its short history has more than made up for a late start. In fact, the first para unit was formed by 370 men of No 2 Commando taking the first full parachute course in 1940. Volunteers formed 2 and 3 Para Battalions (bns) in 1941 and the Parachute Regiment was formally listed with the Infantry of the Line next year. The red, or rather maroon, beret also came in 1942 and the nickname 'Die Rote Teufeln'—the Red Devils—from the Germans for fierce fighting in Tunisia. Wartime casualties in Europe alone totalled 3092 killed and 5386 wounded after only three years of operations. The four-day defence of Arnhem bridge immortalized 2 Para, but it was not to be one of only three of 18 bns to survive postwar disbandment.

The present three bns date from 1948: 4th/6th, 5th (Scottish) and 7th Bns were renumbered and formed into 16th Para Brigade (Bde) and soon based at Aldershot. The bde fell victim with its specialist airborne supporting units to defence cuts in 1977. From 1 Jan 1982 two para bns came under 5 Infantry Bde with a third always in Northern Ireland.

In coming under the command of 5 Bde the paras rediscovered their traditional airborne role. The bde was formed for out-of-area operations and the two para bns would be backed up by 1/7 Duke of Edinburgh's Own Ghurka Rifles. The traditional airborne role in NATO was considered obsolete—with modern air defence weapons the chances of an airborne force flying through the enemy's lines are very slim. For this reason the 3 TA Para bns are part of NATO's front-line infantry forces with a specific commitment to the defence of West Germany and Denmark. In the meantime the regular battalions rotate between Northern Ireland and 5 Bde.

During the 1956 Suez operation 3 Para dropped on and took Gamil Airport, Port Said, and was joined

by sea from Cyprus by 2 Para. Fighting for all three bns followed in Aden and Borneo (SAS roles as well) during 1964-66. Since 1971 a para bn has more often than not been in Northern Ireland.

The paras are organized as ordinary Army infantry bns of 4 rifle, support and HQ companies. The most distinctive special form of equipment is the paratrooper's helmet, the old style in steel and the new in fibre plastic. Also worn is the paratrooper's smock with elasticated cuffs, tail-piece and a single zip. Shoulder flash colours are red for 1 Para, blue for 2 Para, green for 3 Para.

In support are three Territorial Army (TA) bns (4, 10 and 15); 10 Para is the only unit with direct descent from its World War 2 ancestor.

Training for the airborne role is not as frequent as most ranks would like, but the RAF's capability to drop only one bn at a time is being expanded. A regular soldier makes eight jumps (one at night) to earn the parachute wings worn on his right sleeve. Many paras can also boast wings from foreign airborne forces.

In only 24 days of epic action and hard marching the Parachute Regt won 27 awards (including two posthumous VCs and two DSOs) and 29 mentions in dispatches for the retaking of East Falkland, 2 Para being first into Port Stanley. Casualties were 19 killed and 47 wounded.

This paratrooper typifies the British soldier on the islands. He is carrying a normal load for a 'tab' or march. He wears windproof DPM smock and trousers with rank badges on chest and back, steel para helmet with hessian and 'scrimmed' camouflage net, and standard DMS boots with puttees. His weapon is the 7.62mm GPMG or 'jimpy', and he carries some of his belted ammunition over one shoulder. Webbing is normal '58 pattern with the latest model of nylon 'bergen' or rucksack. Sleeping mat is foam plastic—essential in Arctic climates. White spoon is an all-purpose implement, but *never* worn inside the helmet band in battle!

WELSH GUARDS

This is the newest of the foot guards regiments. Within a month of mounting their first King's Guard on St David's Day 1915, they were in the line in France, fighting as part of the Guards Division.

While getting used to ceremonial soldiering in London between the wars, the Welsh Guards also served in the Middle East. The first battalion (1 Bn) was in Gibraltar in 1939 and early the following year joined the British Expeditionary Force (BEF). The second battalion was landed only as the German Panzer spearheads were pushing towards Dunkirk and, with 2 Bn Irish Guards, they held the perimeter at Boulogne until overwhelmed. Meanwhile, 1 Bn was holding up Rommel's 7th Panzer Division.

Such defensive actions, including the defence of the Dunkirk perimeter, where a Welsh Guardsman won the VC, bought time enough for the BEF to be snatched from the beaches.

In June 1942, 1 Bn joined the Thirty Second Guards Brigade (32 Gds Bde) within the newly formed Guards Armoured Division (Gds Arm Div), while 2 Bn formed the divisional reconnaissance (VECCE) regiment.

In July 1944, Gds Arm Div landed in Normandy and was concentrated around Bayeux for the notorious Operation Goodwood. After fierce fighting, the break-out came, for the regiment but it ran into bitter fighting at close quarters. In August, 2 Bn was relieved in the recce role by the Second Household Cavalry Regiment and the Welsh Guards were reconstituted, 1 Bn as mechanised infantry and 2 Bn in Cromwell tanks.

On 3 September 1940, the drive on Brussels began with 32 Gds Bde on the right of the 75-mile axis of advance and with Grenadier, Irish and Coldstream Guards of 5 Gds Arm Bde on the left. The Welsh Guards went flat out, and theirs was the first vehicle to reach the centre of the liberated capital—to an unforgettable welcome.

Triumph turned to frustration at the failure to link with the embattled First Airborne Division at Arnhem, and months of hard fighting lay ahead before the Division reached the Elbe on 5 May.

Peace brought a return to royal and ceremonial duties, as well as commitments to the British Army on the Rhine (BAOR) and service in the conflicts arising from Britain's withdrawal from the colonies.

The Welsh Guards saw action in Palestine, Aden and Cyprus, and returned to Cyprus as part of the UN peacekeeping force in 1975. The regiment has also rotated regularly through Ulster.

On 1 June 1982, 1 Bn Welsh Guards went ashore on the Falklands with the Fifth Infantry Brigade at San Carlos. Without helicopter lift, they were sent forward to Fitzroy by sea, lead elements landing from HMS *Fearless* on the night of 6/7 June. Bad weather delayed the remaining guardsmen of 1 BN, who were embarked aboard *Sir Tristram* and *Sir Galahad*, to be caught by the airstrike on Bluff Cove with the loss of 32 men. men.

A Welsh Guardsman RTO—radio transmission operator—dressed for action in the Falklands. He wears the khaki foot guards beret, with the cap badge blackened in order to eliminate reflection. The combat jacket and trousers are the arctic type, windproofed. The boots, puttees and webbing are all standard. In the nylon rucksack is an A41 Larkspur series radio weighing 45lbs— about the weight of four loaded rifles. The gloves are the combat models first used in Northern Ireland. Wrapping a piece of hessian or similar material round the forearm, and sometimes the butt too, is a common practice designed to cut out reflections from those shiny surfaces.

ROYAL NAVY

The Royal Navy went to war in the South Atlantic little short of a year since a radical and controversial plan for its future had been officially published by the government. It was called 'The Way Forward' and recast the roles and the means of meeting their tasks for Britain's armed forces in the decade to come.

In addition to the commitment to procure the Trident submarine launched ballistic missile, the greatest impact was on the RN's surface forces. In spite of the change of direction recognized in drawing the lessons of the Falklands conflict, such as the decision to retain HMS *Invincible*, the underlying trend is the same.

There will be a reduction of 20 per cent of surface units, and strike power will go above and below the surface—in nuclear powered fleet submarines (SSNs) and in RAF-operated maritime airpower. By 1990 there will be 19 SSNs armed with Sub-Harpoon anti-ship missiles and advanced ASW torpodoes in service. This will add up to a very potent force indeed. They will be joined by a new class of conventional submarines from 1988 onwards and, on current plans, the first of four nuclear powered Trident ballistic missile submarines, each twice as big as the present *Resolution* class Polaris boats, will enter service in the early 1990s.

This Royal Navy anti-aircraft gunner is wearing a blue waterproof anorak and overtrousers and a yellow waistcoat for identification purposes. The anti-flash hood and gloves are worn by everyone on board when a ship goes into action to protect the otherwise exposed flesh of the head and hands from flash burns. The gunner also wears ear protectors to prevent damage to the ear drums from the deafening noise of the gun and he carries an army respirator in a case on his right hip. In this picture the gunner is carrying a clip of anti-aircraft shells. Deploying air power from aircraft carriers and protecting carriers and troopships from enemy air power is one of the major roles of the Royal Navy.

K.Lyles.

Some naval strategists would argue that this change of emphasis reflects the optimum tailoring of forces to meet the NATO role—that is concentration on the NE Atlantic. The RN's primary military tasks here are the ASW support of the US Atlantic fleet, where the *Invincible* class would operate in part under the US umbrella; the defence of Atlantic convoys; submarine search and strike in areas where the Soviet Navy enjoys superiority and where SSNs would be used; and shallow water ASW and mine countermeasures. The RN is concentrating efforts in this last arena on a new generation of small ASW frigates, the Type 23 using towed array sonar and a new medium helicopter, the E101. Air defence for surface units will be provided by the US Phalanx gun system, installed on *Illustrious* before she relieved *Invincible* on the Falklands station, the Sea Wolf missile and the Sea Guard gun system.

The Falklands War demonstrated both the flexibility of sea power and the fact that the RN, in spite of its primary commitment to NATO, can project power out of area at great range and for a prolonged period. Those who wish to see the RN with a higher mix of surface to sub-surface units would point as well to the ever growing global reach of the Soviet Navy (which now includes expanding amphibious forces and organic air power) and the need to be able to meet this threat in strategic areas such as the Gulf of Persia and the Indian Ocean.

The Royal Navy has had a long and distinguished history. And though the daily issue of a tot of rum has gone, along with the firing of broadsides, tarred pigtails and lime eating, some of the senior service's oldest traditions still survive. The RN is still publicity shy and 'silent'. It may not rule the waves but on present planning he RN will still be the world's third most powerful navy in 1990.

ROYAL ARMY MEDICAL CORPS

The medics who went to war in the Falklands belonged to a tradition older than many of the units of the men whose wounds they dressed. But though the story of medicine and the military goes back well into the 17th century it is only comparatively recently that the means of evacuating and treating the wounded—of both sides—has become so mercifully efficient.

And there can be no doubt that the field ambulances of the Royal Army Medical Corps who, among their naval counterparts, fought to save life in the front line of the Falklands campaign were extremely efficient. All but three of the 650 wounded who reached them survived. Some 318 operations were performed at the front, mostly in field hospitals where improvization had to be the order of the day. A frequency generator, for example, was used to relieve pain by electro-acupuncture; and blood in plastic bags was warmed in an old baked bean tin.

But not everything was improvized. A carefully planned sleeping-drug regime allowed pilots to perform 100 hours flying in two weeks—twice the normal maximum—and tackle the 30

Pictured here is a surgeon who is a major in the Royal Army Medical Corps, attached to the paras at the refrigeration plant at Ajax Bay. This makeshift hospital was known as 'the red and green life support machine' as it attended to the red-beretted paras as well as green-beretted commandos. He is wearing an olive green heavy duty woollen pullover with blackened major's crowns on the rank slides and a para brevet on the right sleeve, lightweight trousers, standard issue DMS army boots and puttees. He has a surgeon's protective face mask around his neck and a disposable plastic apron. At Ajax Bay the medics worked under additional dangers. The decision was taken not to paint red crosses on the hospital roof as it was surrounded by logistical stores which were undeniably legitimate military targets.

K.Lyles.

hour round trip from Ascension and back. Experience gained in previous campaigns in the treatment of battle wounds led surgeons to dress them lightly, not closing them for five days, so that any dead flesh or infection missed first time could be tackled easily.

The RAMC traces its ancestry back to 1660 when surgeons were appointed to each regiment under an Inspector General of Hospitals. But it wasn't until the Crimean war, when the commissariat and hospital services all but broke down completely, that the present medical services began to take shape. The Medical Staff Corps was established in 1855, the Army Hospital Corps began providing orderlies in 1875 and the Army Medical School opened in 1860. In 1873 these became the Army Medical Department and regimental hospitals became garrison hospitals, but each regiment now had a Medical Officer. The Army Nursing service was founded in 1881 and finally, on 23 June 1898, the Royal Navy Medical Corps was established, just in time to see action in the Boer War.

The efforts of the corps in the two world wars were aimed as much against disease and infection as battle casualties. And during World War 1 alone, the RAMC itself lost 6873 officers and men.

In World War 2 the RAMC saw action in every theatre which British soldiers fought in and they went into the front line with them by landing craft and parachute. The war also saw the widespread introduction of blood transfusion and, from 1943 onwards, penicillin.

Today, the RAMC, along with the Royal Army Dental Corps and QUARANC nurses, provides the Army's medical service. In peacetime its medical personnel are mainly based on military hospitals. But in wartime they form front line units. Officers and men are combat proficient and trained in combat casualty evacuation using helicopters and tactical vehicles like the Samaritan armoured ambulance.

FLEET AIR ARM

The Fleet Air Arm of the Royal Navy proudly traces its history back to the establishment of the Naval Wing, Royal Flying Corps on 13 May 1912. It was renamed the Royal Naval Air Service a month before the outbreak of war in 1914, and fought independently and effectively as such until integrated into the Royal Air Force in 1918.

Between the wars, naval aviation, afloat and ashore, was the responsibility of the Fleet Air Arm of the RAF, but the service became independent again in May 1939 as the Air Branch of the Royal Navy (called Fleet Air Arm from 1953).

After a proud wartime record, the service began a long period of postwar run down, in spite of the introduction of high performance jet aircraft such as the Phantom and Buccaneer in the 1960s. The 1966 Defence White Paper, however, cancelled the forward building programme and the 1970s brought only a short reprieve for conventional fixed-wing flying. The mighty aircraft carriers HMS *Eagle* and *Ark Royal* were withdrawn in 1971 and 1979 respectively. Meanwhile, a new class of warship had made its debut, the *Invincible* class light carrier, and in June 1979 the RN accepted its first fully operational Sea Harrier: this remarkable VSTOL strike fighter has kept the RN in the forefront of fixed-wing combat flying.

Before the Falklands conflict, the FAA was organized into 16 squadrons with a Fleet Requirement and Direction Training Unit plus a Royal Marine Commando Brigade Squadron composed of four helicopter flights. The FAA operates nearly 250 helicopters in the anti-submarine, transport and now Airborne Early Warning roles. Sea Harriers, with a planned total of 44, provide air defence, reconnaissance and strikepower.

The FAA underwent a rapid transformation in April 1982, going on to a war footing in two days. Equipment itself was improvised, including arming RM Gazelles with rocket pods and fitting Searchwater radar into Sea Kings to provide Airborne Early Warning (AEW) cover. Several squadrons were amalgamated, split or raised from instructors. At RNAS Culdrose, 814 Sqn gave up its Sea Kings to operational units, including 825 Sqn formed on 7 May embarked aboard *Atlantic Causeway* and *QE2*. Lynx Sqns 702 and 815 were split up on destroyers and frigates while Wasps were taken from store for duty on impressed merchant ships. 707 Sqn reformed with Wessex as 848 Sqn and embarked aboard *Olna*, *Regent* and *Atlantic Conveyor*. Another new Wessex unit was 847 Sqn, formed early in May an embarked aboard *Engadine* and *Atlantic Causeway*. Sea Harrier training sqn, 899, gave up some aircraft and pilot instructors to form 809 Sqn on 8 April, embarked aboard *Hermes* and *Invincible*.

The FAA's Sea Harriers proved crucial to the success of the campaign, while Sea Kings provided continuous ASW cover and amphibious lift, and Sea Skua-armed Lynx proved effective against enemy light surface craft. The 28 Sea Harrier jet fighters of 800, 801, 809 and 899 Sqns meanwhile achieved 99 per cent readiness, and flew over 1100 combat air patrol missions and 90 ground attacks. They scored 20 confirmed and three probable kills in air-to-air combat while six were lost, two to ground fire.

The FAA can look forward to the introduction of a new generation helicopter, the EH101, improvements in the proven Sea Harrier, plus a new carrier to join *Illustrious* and *Invincible* in 1985— HMS *Ark Royal*.

This Sea Harrier pilot wears a Mk 10 immersion suit for aircrew flying over water. Under this he wears a G-suit to lessen the effects of centrifugal force on his body during fast manoeuvres. His Mk 36 Helmet (bone dome) is olive green in colour; the visor has a protective cover. He wears the 809 NAS badge on his shoulder.

ROYAL ENGINEERS

Military engineering is as old as the art of war itself, but armies began to develop a recognizable technical 'tail' in the first age of gunpowder. In England the Board of Ordnance emerged in the early 15th century with a Master-General and a permanent staff as the only part of a standing military system. The Board supplied munitions and artillery for fitting out warships or expeditionary forces as the need arose, and for centuries the system just about worked, until 1716 when the artillery and engineers were constituted as separate establishments with an officer corps of engineers (made 'Royal' in 1787). In 1772 a 'Soldier Artificer Company' was formed— originally at Gibraltar—and grew in importance through the Napoleonic wars, being renamed the Corps of Royal Sappers and Miners in 1813. In 1856, following the disastrous logistic breakdowns in the Crimea, the Master General at last lost his personal command of sappers and gunners and the Corps of Royal Engineers was born with officers and men combined in one unit

The Corps has no distinct battle honours because, like the Royal Artillery, it has served in virtually every major British campaign. Its motto 'Ubique' (Everywhere) signifies its breadth of service.

A ribald song of the 1880s described the REs' duties as 'a-digging up of holes, and a-sticking in of poles and a-building of barracks for the soldieree'. But by 1914 the Corps had developed specialities in step with the expanding technology of war. It had formed a signals unit in the 1870s and a balloon detachment in 1883. This became the Air Battalion,

The dirty and dangerous work done by these heroic men is typified by this sapper. He wears a DPM Arctic smock, face veil, denim trousers, DMS boots and puttees, and standard issue wool gloves. The pick is a lightweight infantry model. The plastic Argentine anti-personnel mines could only be detected by prodding for them.

RE, in 1911 and in 1912 formed the basis of the new Royal Flying Corps. The Royal Corps of Signals was formed in 1920.

In World War 1 the RE grew massively in size and importance as the war turned into a long trench siege; their work included signalling and the harrowing and heroic business of tunnelling and mining. The RE grew at peak to well over a quarter of a million men in World War 2 and in 1941 the office of Engineer-in-Chief was created in the War Office. The work of the sappers in the field was immense, from getting harbours working to airfields repair, from minesweeping to running railways and combat assault.

The Royal Engineers today command a very wide range of skills and with the Royal Electrical and Mechanical Engineers (formed in 1943) are responsible for operating highly sophisticated equipment. With BOAR there are four armoured division engineer regiments responsible for working closely with armour in defence and attack, one armoured engineer regt with special skills in minelaying, and one amphibious engineer regt. In the UK there are at present four field engineer regts with six TA engineer regts. In addition there are formations concerned with bomb disposal (the RE is responsible for explosive ordnance disposal below ground), airfield repair sections and a commando squadron.

In the Falklands the Royal Engineers were the real unsung heroes: men from 59 Commando Sqn, 9 Parachute field Sqn, 36 Regt and 3 and 50Fd Sqns repeatedly put their lives on the line, defusing bombs and clearing minefields—often at night and under heavy fire—in advance of the infantry.

On top of that, the sappers built two bridges, laid a Harrier landing strip, built a fuel installation there that dispensed 40,000 gallons of fuel a day, and shot down a Skyhawk. They also moved two tons of mail and 1000 parcels a day and produced 750,000 maps. Unsung heros.

SCOTS GUARDS

The First Regiment of Guards (later the Grenadiers) and the Coldstream Guards nicknamed The Scots Guards the 'kiddies' when they first came on to the strength of the English Army at King James II's camp outside London in 1686. The First and Coldstream had been raised more than 25 years previously, at the Restoration; but in fact, the Scots Guardsmen proudly traced their ancestry still further back, to the Marquess of Argyll's regiment raised in 1642 as Charles II's personal bodyguard for a projected campaign in Ireland.

The motto 'Nemo Me Impune Lacessit' means 'No one molests me with impunity.'

Argyll's regiment fought Cromwell's forces as Charles Stuart's 'Lyfe Guard of Foot', until finally beaten at Worcester in July 1651.

With a Stuart king back on the throne of England and Scotland, in May 1662 the 'Scotch Guards' were raised again to garrison Edinburgh and in 1666 a further seven companies were added. They were now officially established as the Scottish Regiment of Foot Guards, but then came the Act of Union, after which they were re-christened the Third Regiment of Guards Foot Guards, the name they were to retain until 1831.

The Third Regiment, like the Scots Guards of today, combined ceremonial duties with military prowess. From the first battle honour at Namur in 1695 to Waterloo in 1815, they fought the soldiers of France on many battlefields.

The epic stand of the Second Battalion (2 Bn) Grenadiers with 2 Bn Third Guards at the farmhouse of Hougemont proved vital to the victorious outcome of Waterloo and to this day the farmhouse bears a Scots Guards commemorative plaque.

In 1831 the regiment became the Scots Fusiliers Guards and donned bearskins. They got their pipers in 1856 and in 1877 at last became the Scots Guards.

They fought in the Crimea win-

ning two VCs at the Alma; they were at Telel Kebir in 1882; they fought the Mahdi in 1885 and they were involved in the Boer War. The hard lessons learned in South Africa made the British Army of 1914 the most professional in its history—and the Guards, at its cutting edge, suffered accordingly. At the first battle at Ypres, for example, the Scots Guards lost three quarters of its strength. In World War I, the regiment won five VCs on the Western Front.

In World War 2, Scots Guardsmen fought in Norway, the Western Desert and Tunisia—where they specialized in antitank gunnery. They landed at Anzio and Salerno and fought the long slog up Italy. The Third Battalion fought in tanks with the Sixth Grenadier Tank Brigade from Normandy to the Baltic.

Scots Guardsmen (with a high percentage of National Servicemen) fought in the post-war Malayan emergency, in the run-up to Suez in 1956 and in Borneo in 1964-5. More often than not since 1971, a Scots Guard battalion has been rotated through Ulster.

On 1 June 1982, 2 Bn landed at the San Carlos bridgehead with the Fifth Infantry Brigade. On the nights of 5/6 and 6/7 June the Scots were taken by assault ship to Fitzroy, advancing to take Tumbledown Mountain against stiff opposition on the night of 13/14 June.

This Scots Guardsman wears the khaki foot guards beret with a blackened cap badge. He wears the DPM windproof smock common to all Task Force personnel, but instead of DPM trousers he wears the quilted trousers of his 'Mao' suit which have a zip up the side allowing him to pull them on and off without removing his boots. His rifle is the standard British SLR, with a Sight Unit, Infantry, Trilux (SUIT sight) fitted for night fighting. The sight has an illuminated needle which allows a bead to be drawn on the target in darkness. His field dressing is taped to the butt, common practice on the battlefield among both British troops and Argentines.

ROYAL ARTILLERY

The story of artillery in Britain goes back more than five hundred years, almost to the introduction of gunpowder itself. The Board of Ordnance was established in 1414 and the Master-General would release guns as the need arose. For centuries the system creaked along, but during the Jacobite rebellion of 1715 it took so long to mobilize the guns that the fighting was over before any were ready. On 26 May, 1716, two companies of Artillery were created by Royal Warrant, and in 1722 they became the Royal Regiment of Artillery.

The regiment grew in size and importance through the 18th century, but civilian waggons and horses were still hired as necessary. In 1794 the Corps of Captains, Commissaries and Drivers were formed to put them under military discipline, and in fact up to 1918 recruits were still enlisted as 'gunner and driver'.

In 1856 the Master-General lost his separate control and gunners and engineers at last came under the control of the War Office itself. In both world wars the gunners were of course of immense importance, from the trench siege war of 1914-18 to the war of mobile firepower of 1939-45. They have no battle honours, for like the Royal Engineers they have been engaged in practically every British campaign of any importance. The motto 'Ubique' is symbolic of their universal service.

In fact the British Army's artillery today is made up of two components, The Royal Regiment of Artillery and the Royal Horse Artillery (formed in 1793). The RHA is much smaller and acts as an elite corps striving for a special standard of excellence, but of course the RA and RHA function as one. The King's Troop, Royal Horse Artillery, who have a ceremonial function, was founded in 1947.

The Royal Artillery today operates some of the most sophisticated kit in the British Army, including new tactical nuclear weapons and advanced com-

K. Lyles.

puting and communications equipment. Principal systems operated are Lance tactical nuclear missiles, M109 howitzers, M110 howitzers, Abbot SP guns, FH70 towed howitzers, the 105mm light gun, Swingfire long range guided anti-tank missile and Rapier SAMs. The five 105mm batteries of 29 Cmdo and 4th Field Regiment engaged in the Falklands campaign fired nearly 17,500 rounds, and some guns as many as 500 rounds in 24 hours. Rapiers scored 14 confirmed aircraft kills.

Today the RA's main strength is committed to BAOR where there are 9 field regts, 1 heavy regt, 1 missile regt (armed with Lance), 1 anti-tank, 1 locating and 2 air defence regiments with forces assigned to operate with armoured divisions and in their own right.

In the UK normally there are 4 field regts, 1 medium regt, 1 air defence regt, 29 Cdo regt and five major RA regts. The Royal School of Artillery is at Larkhill, Wilts, with four major wings specializing in gunnery, guided weapons, tactics and signals.

Units deployed to the Falklands were 29 Commando Regiment, 8 Battery from 4 Field Regiment, Rapiers of T Battery, 12 Regiment, and Blowpipes of 43 Battery, 32 Guided Weapons Regiment. Later on in the campaign these units were supplemented by TAC HQ of 4 Fd Regt, and that regt's 97 Fd Bty.

The Commando gunners—29 Regiment—are an integral part of 3 Commando Brigade, and are drawn from regular army volunteers who must pass the Green Beret course at Lympstone before joining the regiment.

'Every hard-pressed soldier loves a gunner . . .' The Royal Artillery's contribution to the Falklands campaign was immense. This man wears a non-issue waterproof anorak with DPM trousers and arctic cap. The cap has fur-lined ear-pieces which are only used off duty. He wears black leather Northern Ireland gloves and carries the tool of his trade, a 105mm artillery shell.

BLUES & ROYALS

The men of the Blues and Royals who took their Scorpion light tanks to the Falklands trace their regimental history through two distinguished lines. The regiment was formed on 26 March 1969 by amalgamating the Royal Horse Guards, known almost since their inception as the Blues, and the First Royal Dragoons.

Both the Blues and Royals trace their origins back to the restoration of King Charles II in 1660. Needing to build an army, the King appointed a royalist officer to command an existing regiment of heavy cavalry known as 'The Royal Regiment of Horse'. In 1661 the regiment was reformed under the command of the Earl of Oxford, and they wore distinctive blue uniforms, both in deference to their new commander and in line with their original dress in Cromwell's New Model Army. And when in 1687 the regiment became the Royal Horse Guards, the nickname 'Blues' still stuck. Charles formed another regiment on 21 October 1661 as a troop of horse to be sent to reinforce the garrison of Tangier. On their return, they were reorganized as the 'Royal Dragoons'—later known as the Royals.

In the wars of the 18th century the Blues won battle honours at Dettingen, at Warburg, where in 1760 the regiment's colonel, the Marquis of Granby, led a famous charge against the French—Beaumont and Willems in Holland. And in 1812, the Blues formed the 'Household Brigade' with the First and Second Life Guards to fight in the Peninsular War under Wellington, and later at Waterloo.

Meanwhile the original Tangier

This trooper of the Blues and Royals wears the glass-fibre helmet with built-in radio head set which is issued to all crews of armoured vehicles. The boots are a high-leg model, which suggests the trooper had served in Northern Ireland where this type of boot is standard issue. The 20-litre jerrican contains fuel for his CVR(T). There were never enough cans to go round.

K. Lyles.

Horse became the Royal Regiment of Dragoons in 1690 and the 1st (Royal Dragoons) in 1751. The Royals also won battle honours at Dettingen, Warburg, Beaumont, Willems, in the Peninsula, at Waterloo and in the Crimea and the Boer War. And the Blues won more battle honours at Tel-el-Kebir, in Egypt in 1882 and in South Africa.

In the early stages of World War 1 both cavalry regiments fought in the opening manoeuvre phase, but the subsequent years brought the horse-borne troops bloody frustration, either in the trenches or waiting for the breakthrough which never came.

Nevertheless, both regiments still went to war on horseback in 1939, and fought their first campaign in the Middle East and North Africa as cavalry. The Blues transferred to armoured cars in February 1942 and the Royals made the switch later that same year.

As the Second Household Cavalry Regiment, the Blues fought in north-west Europe in 1944-45 with their armoured cars consistently in the van guard of the Allied advance. After World War 2 the Blues and the Life Guards kept their armoured cars, but normally one regiment was on ceremonial duties while the other was abroad in Germany or Cyprus. The combined regiment of Blues and Royals converted to Chieftain tanks on formation in 1969, switching to the Scorpion in 1974.

Two 14 man Troops went to the Falklands with four Scorpion and four Scimitar light tanks plus a Samson armoured recovery vehicle manned by Royal Electrical and Mechanical Engineers personnel. The vehicles each covered an average 350 miles and performed well despite the soft ground. One Scimitar needed a gearbox change and a Scorpion lost a track to a mine. The Blues and Royals suffered no casualties in the Falklands campaign, and brought back two Argentine Panhard armoured cars as trophies.

GURKHA RIFLES

The first battalion, 7th Gurkha Rifles who fought in the Falklands are part of the British Brigade of Gurkhas who form an historic link with the old Indian Army.

Their predecessors fought in many of the conflicts of British India and the two world wars, and have proved some of Britain's toughest soldiers. The story begins in 1814 when Lt Frederick Young recruited a corps of Gurkha soldiers from the tough hill tribes of Nepal following the war between the army of the East India Company and the independent kingdom.

There is still a depot in Nepal to this day which acts as a recruiting centre for the Gurkhas of the British and Indian armies.

The 7th Gurkhas were founded at Quetta in NW India in 1907, recruited from the 'jats' or clans of eastern Nepal, and in 1908 a second battalion was raised.

2/7th's baptism of fire was in 1915 defending the Suez Canal against the Turks, and their first major battle honour was at Naseriya, advancing through Mesopotamia the same year. The campaign turned to disaster and while 2/7th made an epic stand at Ctesiphon, the battalion went into Turkish captivity following the fall of Kut in April 1916.

The reformed 2/7th was in action again in the Middle East, at Ramadi in 1916 and in Palestine in 1918 where 1/7 also won a battle honour at Sharquat.

Between the wars the 7th soldiered on in Afghanistan and on the NW frontier and a certain Col Slim (later Field Marshal Slim of Burma) became 2/7th's Commander. In May 1941 2/7th were committed to action in Iraq and the

This Gurkha soldier wears a green waterproof over his DPM arctic smock and trousers, and the '43-pattern steel helmet. His black puttees are peculiar to the Gurkha Brigade, as is the kukri hung on his belt. The hilt is covered with non-reflective tape, and the scabbard with DPM camouflage. The image of a professional soldier is completed by the hessian taped to his SLR.

K. Lyles.

occupation of Iran, and as part of 4th Indian Division they fought doggedly in defence of Tobruk. At the end of the siege in June 1941, 2/7th went into captivity.

A new 2/7th was raised and fought in the long slog up Italy—at Cassino, Monte Grillo and the attack on the Gothic Line where they won the battle honour 'Travaleto'. Before returning to India in December 1945 they were briefly in action during the Greek civil war.

1/7th and 3/7th (raised in 1940) were meanwhile on the Burma/Thai border when the Japanese struck at the end of 1941. Fighting rearguard actions all the way to the Indian frontier, they were decimated in the crossing of the Sittang river. The survivors joined together to form one unit and, with 17th Indian Division, they held the road to Imphal through 1943, offering very determined resistance during the last Japanese offensive of March/June 1944, a rifleman winning the VC in the process. 1/7th were at the fore of the fighting during the reconquest of Burma from January 1945.

At the end of World War 2 there were 50 Gurkha bns in the Indian Army. On independence six joined the army of the new state while four regiments of two bns each became the British Brigade of Gurkhas.

For ten years from 1948 the Brigade was in Malaya, and in the 1950s raised the Gurkha Army Service Corps, later Gurkha signal, engineer and transport units.

In the 1960s the Brigade was in action in Sarawak, and in Hong Kong and Brunei where permanent garrisons have since remained.

In 1971 the Brigade was reduced to five bns, totalling 6700 men. The 7th lost its 2 Bn but began an 18-month tour of duty in the UK including mounting Queen's Guard. One Gurkha bn has been permanently in the UK since then with HQ at Church Crookham, near Aldershot.

ARMY AIR CORPS

In the year of the Falklands campaign the Army Air corps fittingly celebrated its silver jubilee, marking 25 years of army flying since the AAC was officially established on 1 September 1957. The new Corps took over light air units already serving within the RAF, the Air Observation Post (AOP) flights flown by officers of the Royal Artillery, and the so-called 'light liaison flights' flown by men of the Glider Pilot Regiment. Both had proud records from World War 2, flying and fighting in such aircraft as AOP Austers and Horsa and Hamilcar gliders. In fact the last OP Auster was retired in 1967.

In 1955 the Joint Experimental Helicopter Unit (JEHU) was set up and the Army began its association with combat rotary wing flying, the principal function of the AAC today. The first helicopter was the Saro Skeeter AOP, which entered service in 1956. When the Corps was set up in 1957 it was to operate unarmed light aircraft and helicopters with an all up weight of less than 4000lb—leaving the RAF with the larger helicopters such as the Whirlwind and the AAC with frail Austers, Skeeters and Bell 47s. Meanwhile, RAF ground crew maintained the aircraft until REME tradesmen were trained to take over and Middle Wallop became AAC school and centre.

The AAC has come a long way since those early days, and is now the biggest single British operator of helicopters. New aircraft such as the Scout, Gazelle and Lynx brought the need for greater continuity in organization, and from 1973 the corps was able to commission, recruit and train officers and men directly. Today the AAC operates some 30 fixed-wing

This Army Air Corps pilot wears the standard aircrew helmet with cloth protecting the visor, and also D.M.S boots and puttees with the now-obsolete olive green denim trousers. His life preserver is standard for all aircrew. Most aircrew carried a side arm for personal protection—this one carried a 9mm Browning.

K.Lyles.

aircraft—Chipmunk trainers and Beaver utility—and 300 Scout, Lynx, Gazelle and Alouette II helicopters.

An AAC regiment consists of two squadrons, one equipped with Gazelles in the reconnaissance role, and the other with SS11-armed Scouts or TOW-armed Lynx in anti-tank role. Five regiments are based in West Germany supporting 1 British Corps, and there are Gazelle liaison flights at Berlin and Wildenrath. An AAC regiment based at Aldergrove and Omagh, in Northern Ireland, controls a permanent squadron with an additional detachment from West Germany and a Beaver flight.

In England, 7 Regt controls a squadron at Netheravon, 3 Flt based at Topcliffe, in Yorkshire, and 2 Flt allotted to NATO's ACE mobile force. Netheravon is the home of 656 Squadron, which took its Gazelles to the Falklands; 657 Squadron is based at Oakington in Cambridgeshire.

The AAC is much more widely dispersed—three Gazelles operate in Belize, 12 Scouts are based in Hong Kong and two of them deploy in Brunei in support of the Gurkhas, enabling the AAC to keep up with jungle flying training. Two three-aircraft Alouette II flights operate in Cyprus, one of them being assigned to the UN. SS11 Scout helicopters based at the British Army Training unit at Suffield, Alberta, provide anti-tank missile training on the ranges.

The AAC Centre at Middle Wallop controls all training and is actively involved with developing new tactical techniques against the fast-changing technology of battlefield helicopters and their weapons. Aircrew to aircraft ratio at 1:1 is low and the Corps is confident that once the Lynx-TOW conversion programme is complete, it will be the single most effective anti-tank force in the British Army, no longer simply providing reconnaissance and utility support, but also fighting in its own right at the very heart of the battlefield.

ROYAL AIR FORCE

The key roles of the RAF today are its contribution to the defence of NATO's centre sector in Germany, the air defence of the United Kingdom and of its Atlantic approaches. Its equipment and tactical doctrine are geared to these tasks, although the flexibility of the RAF's power, backed up by imaginative improvisation, was amply demonstrated in the Falklands conflict.

The RAF had to fight at extreme range, without forward airfields and against aircraft, in theory at least, tactically more suited to fighting for air superiority. It had to provide an 8000 mile logistic train via Ascension and provide not only strategic reconnaissance and maritime surveillance but strategic strikepower as well.

Harrier GR3's, designed for close air support, were fitted with Sidewinder AIM9L air to air missiles and flown as fighters by pilots rapidly retrained in shipboard operations. Nimrod maritime patrol and Hercules transport aircraft were fitted with in-flight refuelling equipment and flown thousands of miles by using Victor tankers, themselves refuelling as they staged aircraft south. Vulcan bombers made raids on Port Stanley airfield and were equipped with Harpoon anti-shipping and Shrike anti-radar missiles. Nimrods were turned into 'fighters' armed with Sidewinder missiles and equipped with Sting Ray torpedoes.

In all the service and its technical backup reacted with immense speed and flexibility to a war it was not designed to fight.

In that fighting the RAF provided the strategic airlift via Ascension with Hercules and chartered civilian aircraft moving 350

This RAF Hercules Air Loadmaster wears zip-up flying boots and olive green flying suit with white leather 'PJI' gloves. He wears an infantryman's DPM smock with epaullette rank slides over his safety harness, an unusual feature. The harness clips to a strong-point in the aircraft, and his headset is connected to the intercom by a jack plug.

K. Lyles.

tons of freight between April and June, while VC10s evacuated casualties from Uruguay. The surviving Chinook helicopter from *Atlantic Conveyor* proved invaluable in support of ground forces carrying way over its design limit without breakdown. The fourteen RAF Harriers were used mainly in the ground attack role using free fall, laser-guided and cluster bombs while Nimrods provided maritime cover. Three GR3s were lost to ground fire.

In spite of the squadron of Phantom interceptors now stationed at RAF Stanley, the RAF's core rationale is of course, remains in Europe. The first of a planned total of 220 Tornado GR1 strike aircraft are just entering service while 165 Tornado F2s will provide air defence. It is planned to modify up to 100 Hawk trainers to carry Sidewinders in the interceptor role and eleven Nimrod AEW3 early warning aircraft, when in service, will provide long range intercept capability.

The front-line capability of RAF Germany will be revitalised by nine Tornado strike squadrons replacing Jaguar and Buccaneers. Two Buccaneer sqns will receive avionics modernisation and be armed with Sea Eagle anti-ship missiles to continue in the maritime strike role.

Harrier GR3s in Germany will be eventually replaced by sixty Harrier GR5s, RAF versions of the UK/US Harrier development called the AV-8B. The RAF is also looking to a lightweight dogfighter for the 1990s.

The RAF is organised in three commands, Strike Command which embraces Bomber (No 1) Fighter (No 11) and Maritime (No 18) Groups plus the general purpose group (No 38) which is also responsible for the RAF Regiment, tasked with airfield defence and equipped with Rapier SAMS. Support Command provided material and training support while the Third Command is RAF Germany itself, which is a component of NATO's 2nd Allied Tactical Air Force.

SPECIAL AIR SERVICE

The Special Air Service is as shadowy and secretive as submarines—and it excites as much, if not more curiosity. Its unique method of selection and training, and its operational role, have ensured in its 40 year history a continuity of this 'specialness'. Since 1941 men of the SAS have seen service in over 30 different theatres of war, been in virtually continuous action, and spawned many foreign equivalents.

The Special Air Service was born in 1940-41 at the lowest ebb of the war for Britain. Churchill had called for the setting up of raiding units able to hit back against Occupied Europe. By 1940, fledgling commando and paratroop units had been formed. In November, 2 Commando was renamed, '11 Special Air Service Battalion' and in February 1941 'X'-Troop from 11 SAS Bn made the first-ever British para-commando raid against an aqueduct in Italy. It was then that 11 SAS Bn became the core of 1st Parachute Battalion.

Meanwhile, in the Middle East, Colonel David Stirling had persuaded the C-in-C to form a small force of para-commandos to wage irregular warfare behind the enemy lines in North Africa.

By 1942, the SAS had adopted its badge, the winged dagger, and been joined by French and Greek units and a Special Boat Section.

SAS regiments were next committed in Italy, then in January 1944, the SAS Brigade was formed with two British and Commonwealth Regiments—1 and 2; two French regiments—3 and 4. A Belgian Squadron, 5 SAS, was added later. Already the Brigade

This soldier in the SAS wears a DPM windproof anorak (not the arctic smock) and civilian hiking boots. SAS operational headgear is a matter of choice; this man wears a rolled-up balaclava. His belt order is a mixture of '44 and '58 pattern webbing with US ammunition pouches for his M16 Rifle. A smoke grenade hangs from his belt and the working parts of his rifle lie in a DPM cap.

K. Lyles.

had come a long way from the original concept of small raiding parties and was now to be committed as conventional airborne troops. After D-Day, however, the old style was revived and the SAS carried out a series of raids behind enemy lines, often operating with resistance groups.

In October 1945, the British SAS regiments were disbanded and the French and Belgian ones absorbed into their own armies. But in 1947, a territorial regiment was raised, based on the old London volunteer regiment, The Artists Rifles.

It was thought originally that the SAS were committed to a role on any future European battlefield, but the subsequent SAS story has been one of fighting in the far-flung conflicts of colonial withdrawal. In fact, it was during the Malayan emergency that a special unit called the 'Malayan Scouts' operated with a squadron of 21 SAS and from this unit was spawned 22 SAS, the regular unit in today's army.

From the early '70s, one squadron of the SAS began to concentrate its training on specialist anti-terrorist operations, drawing an ever-tighter veil of secrecy over operations and methods. The successful release of hostages at the Iran Embassy siege of May 1980 revealed how skilled they had become.

Today, the SAS consists of 22 Regiment based at Hereford and 21 and 23 Territorial Regiments. Both regular and territorial formations require the same high standard of recruit. But only 22 Regiment is trained in anti-terrorist operations and in turn it only recruits from the Army itself. The headquarters of 22 SAS controls a training wing, an administrative wing, a signals squadron and a number of so-called 'sabre squadrons'.

In the South Atlantic, the SAS helped in the recapture of South Georgia, in clandestine reconnaissance and airfield attack and in operations on the Argentine mainland reporting aircraft movements.

Falklands Heroes: Roll of Honour

This section records those whose bravery won them, and their units, honours for actions on the field of battle plus a detailed map of the theatre of operations. The participating Regiments are listed, and there is an examination of that peculiarly British institution, stamp collecting — and the miniature war that grew up around these internationally recognised symbols of sovereignty.

Falklands Armoury concludes with a record of the most important and irreplaceable part of any nation's resource — those men and women who were prepared to lay down their lives for it.

ROYAL NAVY, ROYAL MARINES, ROYAL FLEET AUXILIARY AND MERCHANT NAVY

Distinguished Service Order

Commodore Samuel Clark DUNLOP CBE, Royal Fleet Auxiliary; Captain Michael Ernest BARROW, Royal Navy; Captain John Jeremy BLACK MBE, Royal Navy; Captain William Robert CANNING, Royal Navy; Captain John Francis COWARD, Royal Navy; Captain Peter George Valentine DINGEMANS, Royal Navy;

From the citation for
Captain E S J Larken's DSO
'During air attacks, he conducted his ship's defence personally from the exposed gun direction platform and, in so doing, was an inspiring example of personal bravery to his men'.

Captain Edmund Shackleton Jeremy LARKEN, Royal Navy; Captain Christopher Hope LAYMAN MVO, Royal Navy; Captain Linley Eric MIDDLETON ADC, Royal Navy; Captain Philip Jeremy George ROBERTS, Royal Fleet Auxiliary; Captain B G Y YOUNG, Royal Navy; Lieutenant Colonel Nicholas Francis VAUX, Royal Marines; Lieutenant Colonel Andrew Francis WHITEHEAD, Royal Marines; Commander Christopher Louis WREFORD-BROWN, Royal Navy; Lieutenant Commander Brian Frederick DUTTON QGM, Royal Navy; Lieutenant Commander Ian STANLEY, Royal Navy.

Captain Brian Young, DSO, of HMS Antrim

Distinguished Service Cross, Posthumous

Captain Ian Harry NORTH, Merchant Navy; Lieutenant Commander Gordon Walter James BATT, Royal Navy; Lieutenant Commander John Stuart WOODHEAD, Royal Navy; Lieutenant Commander John Murray SEPHTON, Royal Navy.

Distinguished Service Cross

Captain George Robert GREEN, Royal Fleet Auxiliary; Captain David Everett LAWRENCE, Royal Fleet Auxiliary; Captain Anthony Francis PITT, Royal Fleet Auxiliary; Commander Paul Jeffrey

ROLL OF *HONOUR*

In practice, medals and decorations awarded to British armed service personnel for gallantry on active service fall into a number of grades. Each includes decorations for roughly comparable acts by men of the different services; and each includes two parallel sequences of awards, one for officers and one for enlisted ranks: traditionally separate sequences which do not, however, imply any difference in merit: the RAF officer's Distinguished Flying Cross for instance, and the RAF sergeant's Distinguished Flying Medal denote no difference in bravery whatsoever.

The grades are first and foremost, the Victoria Cross; second, the Distinguished Service Order, Distinguished Conduct Medal and Conspicuous Gallantry Medal; third, the Military Cross, Distinguished Service Cross, Distinguished Flying Cross, Military Medal, Distinguished Service Medal and Distinguished Flying Medal; fourth, the Air Force Cross and Air Force Medal. A fifth category comprises awards and honours such as Mention in Dispatches which are not gallantry medals as such, but carry special insignia. Two further categories, outside this sequence, include the George Medal and the civil and military divisions of the Order of the British Empire.

BOOTHERSTONE, Royal Navy; Commander Christopher John Sinclair CRAIG, Royal Navy; Commander Anthony MORTON, Royal Navy; Commander Nicholas John TOBIN, Royal Navy; Commander Nigel David WARD AFC, Royal Navy; Commander Alan William John West, Royal Navy; Lieutenant Commander Andrew Donaldson AULD, Royal Navy; Lieutenant Commander Michael Dennison BOOTH, Royal Navy; Lieutenant Commander Hugh Sinclair CLARK, Royal Navy; Lieutenant Commander John Anthony ELLERBECK, Royal Navy;

From the citation for
Lieutenant Commander
N W Thomas's DSC
'On one occasion, he shot down one of a wave of four Skyhawks and in the ensuing dog-fight in cloud and when his remaining missile indicated that it had acquired a target, he showed great coolness in holding his fire until he was able to confirm that it had in fact detected his wingman rather than the enemy, so preventing a tragic accident'.

VICTORIA CROSS

Instituted in the Crimean War, the VC is awarded to officers and men alike as the supreme decoration for gallantry under fire. The simple bronze cross—originally cast from the metal of captured Russian cannon—is awarded only for a specific act in battle, performed in complete disregard for personal safety. About a quarter of all awards have been posthumous; and between 1945 and April 1982 only four awards had been made to British service personnel. The claret-coloured ribbon is unique and unmistakable.

Lieutenant Commander Hugh John LOMAS, Royal Navy; Lieutenant Commander Neil Wynell THOMAS, Royal Navy; Lieutenant Commander Simon Clive THORNEWILL, Royal Navy; Lieutenant Alan Reginald Courtenay BENNET, Royal Navy; Lieutenant Nigel Arthur BRUEN, Royal Navy; Lieutenant Richard HUTCHINGS, Royal Marines; Acting Lieutenant Keith Paul MILLS, Royal

Lt Richard Hutchings, DSC, his wife and two sons

From the citation for
Sub Lieutenant
P T Morgan's DSC
'On two occasions he dived into the flooded forward magazine, in the knowledge that in addition to the hazards posed by twisted and jagged metal, there was an unexploded bomb in the compartment amongst damaged ordnance'.

Marines; Lieutenant Nigel John NORTH, Royal Navy; Lieutenant Stephen Robert THOMAS, Royal Navy; Sub Lieutenant Peter Thomas MORGAN, Royal Navy; Fleet Chief Petty Officer (Diver) Michael George FELLOWS BEM.

Military Cross

Major Charles Peter CAMERON, Royal Marines; Captain Peter Murray BABBINGTON, Royal Marines; Lieutenant Clive Idris DYTOR, Royal Marines; Lieutenant Christopher FOX, Royal Marines; Lieutenant David James STEWART, Royal Marines.

Distinguished Flying Cross, Posthumous

Lieutenant Richard James NUNN, Royal Marines.

Distinguished Flying Cross

Captain Jeffrey Peter NIBLETT, Royal Marines.

Air Force Cross

Lieutenant Commander Douglas John Smiley SQUIER, Royal Navy; Lieutenant Commander Ralph John Stuart WYKES-SNEYD, Royal Navy.

Distinguished Conduct Medal

Corporal Julian BURDETT, Royal Marines.

George Medal, Posthumous
Second Engineer Officer Paul Anderson HENRY, Royal Fleet Auxiliary.

George Medal
Able Seaman (Radar) John Edward DILLON.

From the citation for Able seaman (Radar) J E Dillon's GM
'He extricated himself and despite pain from a large shrapnel wound in his back attempted unsuccessfully to free a man pinned down by a girder across his neck. He then made his way through the smoke towards a further man calling for help, whom he found trapped under heavy metal girders, bleeding from head and face wounds and with his left hand severely damaged. After several attempts between which he had to drop to the deck to get breathable air, AB/R/ Dillon succeeded in raising the debris sufficiently to allow the man to drag himself free. AB/R/ Dillon's antiflash hood had been ripped off in the explosion, so afforded him no protection from the heat, and his left ear was burned'.

Distinguished Service Medal, Posthumous
Petty Officer Marine Engineering Mechanic (M) David Richard BRIGGS; Acting Corporal Aircrewman Michael David LOVE, Royal Marines.

Distinguished Service Medal
Colour Sergeant Michael James FRANCIS, Royal Marines; Sergeant Peter James LEACH, Royal Marines; Chief Marine Engineering Mechanic (M) Michael David TOWNSEND; Chief Petty Officer (Diver) Graham Michael TROTTER; Chief Petty Officer Aircrewman Malcolm John TUPPER; Petty Officer John Steven LEAKE; Sergeant William John LESLIE, Royal Marines; Acting Petty Officer (Sonar) (SM) Graham John Robert LIBBY; Leading Aircrewman Peter Blair IMRIE; Leading Seaman (Radar) Jeffrey David WARREN.

Military Medal
Sergeant Thomas COLLINGS, Royal Marines; Sergeant Michael COLLINS, Royal Marines; Sergeant Joseph Desmond WASSELL, Royal Marines; Corporal Michael ECCLES, Royal Marines; Corporal David HUNT, Royal Marines; Corporal Stephen Charles NEWLAND, Royal Marines; Corporal Harry SIDDALL, Royal Marines; Corporal Chrystie Nigel Hanslip WARD, Royal Marines; Acting Corporal Andrew Ronald BISHOP, Royal Marines; Marine Gary William MARSHAL, Royal Marines.

Distinguished Flying Medal
Sergeant William Christopher O'BRIEN, Royal Marines.

Queen's Gallantry Medal, Posthumous
Acting Colour Sergeant Brian JOHNSTON, Royal Marines.

1 DISTINGUISHED CONDUCT MEDAL
2 CONSPICUOUS GALLANTRY MEDAL
These are the equivalent decorations, for enlisted ranks, of the officers' DSO. They date from the Crimean War. The DCM is for Army personnel, the CGM and more recent CGM (Flying) for Royal Navy, Royal Marines, Merchant Navy and Royal Air Force personnel. They are rarely awarded and highly prized, and many awards have been for acts which were 'borderline' VC recommendations. The ribbons are in the traditional colours of the three services: red and blue, white and blue, and light and dark blue.

Queen's Gallantry Medal
Chief Engineer Officer Charles Kenneth Arthur ADAMS, Royal Fleet Auxiliary; Lieutenant John Kenneth BOUGHTON, Royal Navy; Lieutenant Philip James SHELDON, Royal Navy; Third Officer Andrew GUDGEON, Royal Fleet Auxiliary; Third Engineer Brian Robert WILLIAMS, Merchand Navy; Marine Engineering Artificer (M) 1st Class Kenneth ENTICKNAPP; Petty Officer Medical Assistant Gerald Andrew MEAGER.

From the citation for Third Officer A Gudgeon's QGM
'On two occasions during this time [the campaign] he showed great courage in risking his life in order to save others. When HMS Antelope blew up and caught fire in San Carlos Water, he volunteered to cox the crash boat to pick up survivors. This he did knowing that HMS Antelope had a second unexploded bomb on board. Despite the fire spreading rapidly, he carried out the rescue of several survivors.'

DISTINGUISHED SERVICE ORDER
Instituted in 1886, for officers 'who had rendered meritorious or distinguished service in war', this honour is awarded only to officers who have been mentioned in dispatches for their conduct of active operations. Typically, it is the reward for a unit commander of one of the fighting services who shows especially skilled and courageous leadership qualities in action. The blue-edged red ribbon of the beautiful enamelled cross recalls the ribbons of the Army Gold Cross and other medals of the Napoleonic Wars.

Chief Officer Peter Hill, Mention in Dispatches

Mention in Despatches
Chief Officer John Keith BROCKLEHURST, Merchant Navy; Commander Robert Duncan FERGUSON, Royal Navy; Chief Officer Peter Ferris HILL, Royal Fleet Auxiliary; Major Peter Ralph LAMB, Royal Marines; Commander Roger Charles LANE-NOTT, Royal Navy; Commander Thomas Maitland Le MARCHAND, Royal Navy; Major Michael John NORMAN, Royal Marines; Major David Anthony PENNEFATHER, Royal Marines; Chief Engineer James Mailer STEWART, Merchant Navy; Commander James Bradley TAYLOR, Royal Navy; Commander Bryan Geoffrey TELFER, Royal Navy; Major Rupert Cornelius VAN DER HORST, Royal Marines; Lieutenant Commander Michael Stephen BLISSETT, Royal Navy; Lieutenant Commander Barry William BRYANT, Royal Navy; Lieutenant Commander Robert Gerwyn BURROWS, Royal Navy; Lieutenant Commander John Sydney Maurice CHANDLER, Royal Navy; Lieutenant Commander John Normanton CLARK, Royal Navy; Lieutenant Commander William Edgar HURST, Royal Navy; Lieutenant Commander John PARRY, Royal Navy; Captain Michael Anthoy Falle COLE, Royal Marines; Lieutenant Commander Gervais Richard Arthur CORYTON, Royal Navy; Lieutenant Commander Rodney Vincent FREDERIKSEN, Royal Navy; Lieutenant Commander David Gordon GARWOOD, Royal Navy; Lieutenant Commander Andrew Clive GWILLIAM, Royal Navy; Lieutenant Commander Lon Stuart Grant HULME, Royal Navy; Lieutenant Commander Ian INSKIP, Royal Navy; Lieutenant Commander Robin Sean Gerald KENT, Royal Navy; Lieutenant Commander John Andrew LISTER, Royal Navy; Lieutenant Commander Iain Bruce MACKAY, Royal Navy; Lieutenant Commander Clive Ronald WELLESLEY MORRELL, Royal Navy; Lieutenant Commander Kenneth Maclean NAPIER, Royal Navy; Captain Andew Bennett NEWCOMBE, Royal Marines; Lieutenant Commander Michael John O'CONNELL, Royal Navy; Captain Eugene Joseph O'KANE, Royal Marines; Captain Andrew Robert PILLAR, Royal Marines; Captain Nicholas Ernest POUNDS, Royal Marines; Lieutenant Commander Alvin Arnold RICH, Royal Navy; Lieutenant Commander Robert Ernald WILKINSON, Royal Navy; Lieutenant Philip James BARBER, Royal Navy; Lieutenant Nicholas Abraham Marsh BUTLER, Royal Navy; Lieutenant Christian Thomas Gordon CAROE, Royal Marines; Lieutenant Christopher Hugh Trevor CLAYTON, Royal Navy; Lieutenant Ronald Lindsay CRAWFORD, Royal Marines; Lieutenant William Alan CURTIS, Royal Navy (Posthumous); Lieutenant Andrew John EBBENS, Royal Marines; Lieutenant William James Truman FEWTRELL, Royal Navy; Lieutenant Fraser HADDOW, Royal Marines; Lieutenant Robert Ian HORTON, Royal Navy; Lieutenant Herbert John LEDINGHAM, Royal Navy; Lieutenant David Anthony LORD, Royal Navy; Lieutenant Peter Charles MANLEY, Royal Navy; Lieutenant Andrew Nevill McHARG, Royal Navy; Lieutenant John Andrew Gordon MILLER, Royal Navy; Lieutenant Paul Graham MILLER, Royal Navy; Lieutenant Andrew Gerald MOLL, Royal Navy; Lieutenant Richard John ORMSHAW, Royal Navy; Lieutenant Christopher Laurence PALMER, Royal Navy; Lieutenant Roland Frederick

PLAYFORD, Royal Marines; Lieutenant Christopher James POLLARD, Royal Navy; Lieutenant Anthony PRINGLE, Royal Navy; Lieutenant Peter Iain Mackay RAINEY, Royal Navy; Lieutenant Frederick William ROBERTSON, Royal Navy; Lieutenant Robin Edgar John SLEEMAN, Royal Navy; Lieutenant David Alexander Bereton SMITH, Royal Navy; Lieutenant Nicholas TAYLOR, Royal Navy (Posthumous); Lieutenant Christopher TODHUNTER, Royal Navy; Lieutenant D A H WELLS, Royal Navy; Sub Lieutenant Richard John BARKER, Royal Navy; Sub Lieutenant Stewart Greig COOPER, Royal Navy; Sub Lieutenant Richard Charles EMLY, Royal Navy (Posthumous); Sub Lieutenant David Edgar GRAHAM, Royal Navy; Sub Lieutenant Paul John HUMPHREYS, Royal Navy; Midshipman Mark Thomas FLETCHER, Royal Navy; Fleet Chief Marine Engineer Artificer (P) Ernest Malcolm UREN; Warrant Officer Class 2 Robert John BROWN, Royal Marines; Warrant Officer Class 2 Adrian Spencer ROBINSON, Royal Marines; Chief Air Engineering Artificer (M) Richard John BENTLEY; Marine Engineering Artificer (H) 1st Class Derek Adrian BUGDEN; Colour Sergeant Barrie DAVIES, Royal Marines; Weapon Engineering Artificer 1st Class Anthony Charles EDDINGTON (Posthumous); Chief Marine Engineering Artificer (H) Keith William GOLDIE; Chief Petty Officer (Ops) (M) Eric GRAHAM; Chief Petty Officer (Diver) Brian Thomas GUNNELL; Marine Engineering Artificer (H) 1st Class Peter Gerhard JAKEMAN; Marine Engineering Artificer (M) 1st Class Kevin MARTIN; Marine Engineering Mechanician (M) 1st Class Timothy MILES; Marine Engineering Artificer (M) 1st Class Stephen Derek MITCHELL; Weapon Engineering Mechanician 1st Class Peter Robert MOIR; Marine Engineering Mechanician (M) 1st Class Hugh Bromley PORTER; Marine Engineering Mechanician 1st Class Alan Gordon SIDDLE; Chief Marine Engineering Mechanic Tyrone George SMITH; Marine Engineering Artificer (M) 1st Class Simon Patrick TARABELLA; Acting Chief Weapon Engineering Mechanician Michael Gordon TILL (Posthumous); Marine Engineering Mechanician (L) 1st Class William Geoffrey WADDINGTON; Colour Sergeant Everett YOUNG, Royal Marines; Petty Officer Aircrewman Alan ASHDOWN; Petty Officer Aircrewman John Arthur BALLS, BEM; Petty Officer Aircrewman David Brian FITZGERALD; Sergeant Peter BEEVERS, Royal Marines; Sergeant Ian William BRICE, Royal Marines; Sergeant Edward Lindsay BUCKLEY, Royal Marines; Sergeant Brian Gordon BURGESS, Royal Marines; Petty Officer Aircrewman Richard BURNETT; Sergeant Edgar Robert CANDLISH, Royal Marines; Sergeant Robert Terence COOPER, Royal Marines; Sergeant Graham DANCE, Royal Marines; Sergeant Colin Charles DE LA COUR, QGM, Royal Marines; Sergeant Brian DOLIVERA, Royal Marines; Petty Officer Marine Engineering Mechanic (M) John Richard ELLIS; Sergeant Andrew Peter EVANS, Royal Marines (Posthumous); Sergeant Ian David FISK,

1 MILITARY CROSS

2 DISTINGUISHED SERVICE CROSS

3 DISTINGUISHED FLYING CROSS

These decorations were instituted during World War 1 for junior officers of the Army, Royal Navy and Marines and Royal Air Force respectively. The MC and DSC are reserved for officers up to the rank of major and commander respectively. The awards are for 'gallant and distinguished services in action' and in the case of the DFC it is only awarded for acts performed while flying on active operations against the enemy. The typical award is to a junior officer who leads his men in battle with exceptional courage and resourcefulness. Aircrew officers may be awarded the DFC either for consistent skill and courage over a number of missions, or for single missions. The purple and white of the MC ribbon are repeated in that of the DFC, in the multi-striped arrangement traditional for the RAF; the DSC retains the Royal Navy's austere blue and white.

1 MILITARY MEDAL

2 DISTINGUISHED SERVICE MEDAL

REVERSE

3 DISTINGUISHED FLYING MEDAL

REVERSE

These are the equivalent decorations to the MC, DSC and DFC for enlisted ranks, and also date from World War 1. The memorandum instituting the DSM in 1914 best sums up the conditions of award of all three medals: it was for men who '. . . show themselves to the fore in action, and set an example of bravery and resource under fire'. Awards are made both for individual acts of courage, and for leadership in battle by NCOs. The MM ribbon is in national colours; both the DSM and the DFM ribbons are slight variations on the design of the equivalent officers' awards.

Royal Marines; Weapons Engineering Artificer 2nd Class Jonathan Martin Charles FOY; Sergeant David Keith HADLOW, Royal Marines; Sergeant Kevin Michael JAMES QGM, Royal Marines; Petty Officer (Missile) Hugh JONES; Marine Engineering Artificer 2nd Class David John LEANING; Sergeant William David Paul LEWIS, Royal Marines; Sergeant Mitchell McINTYRE, Royal Marines; Sergeant Henry Frederick NAPIER, Royal Marines; Petty Officer (Rader) Jack PEARSON; Petty Officer Air Engineering Mechanic (M) Stuart RAINSBURY; Acting Petty Officer Marine Engineering Mechanic (M) David Morgan Kerlin ROSS; Sergeant Thomas Arthur SANDS, Royal Marines; Sergeant William John STOCKS, Royal Marines; Sergeant Christopher Ralph STONE, Royal Marines; Petty Officer Aircrewman Colin William TATTERSALL; Weapon Engineering Mechanician 2nd Class Barry James WALLIS (Posthumous); Sergeant Robert David WRIGHT, Royal Marines; Acting Leading Medical Assistant George BLACK; Acting Leading Marine Engineering Mechanic (M) Craig Robert BOSWELL; Corporal Christopher John Graham BROWN, Royal Marines; Corporal Gordon COOKE, Royal Marines; Leading Seaman (Missile) Robert Marshall GOULD; Leading Aircrewman James Andrew HARPER; Acting Leading Marine Engineering Mechanic (M) Stanley William HATHAWAY; Leading Radio Operator (Tactical) Roderick John HUTCHESON; Leading Seaman (Diver) Phillip Martin KEARNS; Corporal Thomas William McMAHON, Royal Marines; Leading Aircrewman Ian ROBERTSON; Leading Seaman (Diver) Charles Anthony SMITHARD; Leading Seaman (Diver) Anthony Savour THOMPSON; Leading Aircrewman Stephen William WRIGHT; Leading Medical Assistant Paul YOUNGMAN; Radio Operator (Tactical) 1st Class Richard John ASH; Lance Corporal Peter William BOORN, Royal Marines; Able Seaman (Missile) Nicholas Scott BROTHERTON; Marine Engineering Mechanic (M) 1st Class Lee CARTWRIGHT; Marine Engineering Mechanic (M) 1st Class Michael Lindsay CHIPLEN; Able Seaman (Missile) Andrew COPPELL; Marine Engineering Mechanic (M) 1st Class Christopher CROWLEY; Marine Engineering Mechanic (M) 1st Class David John EDWARDS; Lance Corporal Barry GILBERT, Royal Marines; Able Seaman (Missile)

Stephen INGLEBY; Able Seaman (Radar) Mark Stanley LEACH; Medical Assistant Michael NICELY; Marine Engineering Mechanic (M) 1st Class David John SERRELL; Marine Engineering Mechanic (M) 1st Class Alan STEWART; Able Seaman (Diver) David WALTON; Marine Robert BAINBRIDGE, Royal Marines; Marine Nicholas John BARNETT, Royal Marines; Marine David Stanley COMBES, Royal Marines; Marine Gary CUTHELL, Royal Marines; Marine Leslie DANIELS, Royal Marines; Marine Stephen DUGGAN, Royal Marines; Marine Leonard John GOLDSMITH, Royal Marines; Marine Graham HODKINSON, Royal Marines; Marine Mark Andrew NEAT, Royal Marines; Marine Geoffrey NORDASS, Royal Marines; Marine David Lloyd O'CONNOR, Royal Marines; Marine Christopher James SCRIVENER, Royal Marines; Marine John STONESTREET, Royal Marines; Marine Ricky Shaun STRANGE, Royal Marines; Marine Perry THOMASON, Royal Marines; Seaman (OPS) Douglas James WHILD; Marine Paul Kevin WILSON, Royal Marines.

Queen's Commendation for Brave Conduct

Second Officer Ian POVEY, Royal Fleet Auxiliary; Chief Marine Engineering Mechanic (L) Alan Frank FAZACKERLEY; Chief Weapon Engineering Mechanic (R) William RUMSEY; Weapon Engineering Mechanic (R) 1st Class John Richard JESSON; Marine Engineering Mechanician (M) 1st Class Thomas Arthur SUTTON; Acting Colour Sergeant David Alfred WATKINS, Royal Marines; Petty Officer Class 2 Boleslaw CZARNECKI, Merchant Navy; Petty Officer Weapon Engineering Mechanic (R) Graeme John LOWDEN; Radio Operator (Tactical) 1st Class David Frederick SULLIVAN; Marine Paul Anthony CRUDEN, Royal Marines.

ARMY

Victoria Cross (Posthumous)

Lieutenant Colonel Herbert JONES OBE, The Parachute Regiment; Sergeant Ian John McKAY, The Parachute Regiment.

Distinguished Service Order

Major Cedric Norman George DELVES, The Devonshire and Dorset Regiment; Major Christopher Patrick Benedict KEEBLE, The Parachute Regiment; Lieutenant Colonel Hew William Royston PIKE MBE, The Parachute Regiment; Lieutenant Colonel Michael Ian Eldon SCOTT, Scots Guards.

From the citation for
Major C N G Delves's DSO

'Following the successful establishment of the beachhead in San Carlos Water, Major Delves took his SAS Squadron 40 miles behind the enemy lines and established a position overlooking the main enemy stronghold in Port Stanley where at least 7000 troops were known to be based. By a series of swift operations, skilful concealment and lightning attacks against patrols sent out to find him, he was able to secure a sufficiently firm hold on the area after ten days for the conventional forces to be brought in'.

Distinguished Service Cross

Warrant Officer Class 2 John Henry PHILLIPS, Corps of Royal Engineers.

Military Cross, Posthumous

Captain Gavin John HAMILTON, The Green Howards (Alexandra, Princess of Wales' own Yorkshire Regiment).

Military Cross

Major Michael Hugh ARGUE, The Parachute Regiment; Captain Timothy William BURLS, The Parachute Regiment; Major David Alan COLLETT, The Parachute Regiment; Lieutenant Colin Spencer CONNOR, The Parachute Regiment; Major John Harry CROSLAND, The Parachute Regiment; Major Charles Dair FARRAR-HOCKLEY, The Parachute Regiment; Major John Panton KISZELY, Scots Guards; Lieutenant Robert Alasdair Davidson LAWRENCE, Scots Guards; Captain William Andrew McCRACKEN, Royal Regiment of Artillery; Captain Aldwin James Glendinning WIGHT, Welsh Guards.

From the citation for
Major J P Kiszely's MC

'Under fire and with a complete disregard for his own safety, he led a group of his men up a gully towards the enemy. Despite men falling wounded beside him he continued his charge, throwing grenades as he went. Arriving on the enemy position, he killed two enemy with his rifle and a third with his bayonet. His courageous action forced the surrender of the remainder. His was the culminating action in the Battalion successfully seizing its objective'.

Distinguished Flying Cross

Captain Samuel Murray DRENNAN, Army Air Corps; Captain John Gordon GREENHALGH, Royal Corps of Transport.

Distinguished Conduct Medal, Posthumous

Private Stephen ILLINGSWORTH, The Parachute Regiment; Guardsman James Boyle Curran REYNOLDS, Scots Guards.

Distinguished Conduct Medal

Corporal David ABOLS, The Parachute Regiment; Staff Sergeant Brian FAULKNER, The Parachute Regiment; Sergeant John Clifford MEREDITH, The Parachute Regiment; Warrant Officer Class 2 William NICOL, Scots Guards; Sergeant John Stuart PETTINGER, The Parachute Regiment.

From the citation for
Staff Sergeant B Faulkner's DCM

'He never faltered, setting a magnificent personal example of courage and competence, that was well beyond anything that could reasonably be expected. One burst of shellfire left him concussed, but he swiftly returned to his duties. One minute he could be seen consoling young soldiers, severely distressed by the experience of losing their comrades... then yet again tending for the casualties themselves'.

Conspicuous Gallantry Medal, Posthumous

Staff Sergeant James PRESCOTT, Corps of Royal Engineers.

Military Medal, Posthumous

Private Richard John de Mansfield ABSOLON, The Parachute Regiment; Lance Corporal Gary David BINGLEY, The Parachute Regiment.

Left to right: L/cpl Bentley, Sgt Barrett, Pte Grayling and MMs

Military Medal

Corporal Ian Phillip BAILEY, The Parachute Regiment; Lance Corporal Stephen Alan BARDSLEY, The Parachute Regiment; Sergeant Terence Irving BARRETT, The Parachute Regiment; Lance Corporal Martin William Lester BENTLEY, The Parachute Regiment; Sergeant Derrick Sidney BOULTBY, Royal Corps of Transport; Corporal Trevor BROOKES, Royal Corps of Signals; Corporal Thomas James CAMP, The Parachute Regiment; Private Graham Stuart CARTER, The Parachute Regiment; Guardsman Stephen Mark CHAPMAN, Welsh Guards; Corporal John Anthony FORAN, Corps of

---

From the citation for
Corporal J A Foran's MM

'During the assault, Corporal Foran, Royal Engineers, led a patrol through an unmarked enemy minefield to assault an enemy position. The patrol came under heavy fire, a burst from a machine gun killing two men. A further two men were wounded by exploding mines.

Without hesitation and completely disregarding his own safety, Corporal Foran re-entered the minefield and cleared a path to his injured colleagues. Having treated them he cleared a route back out of the minefield, enabling the casualties to be evacuated'.

L/Cpl Dale Loveridge, MM, and his fiancée Barbara

Royal Engineers; Sergeant Desmond FULLER, The Parachute Regiment; Private Barry James GRAYLING, The Parachute Regiment; Corporal Thomas William HARLEY, The Parachute Regiment; Bombardier Edward Morris HOLT, Royal Regiment of Artillery; Sergeant Robert White JACKSON, Scots Guards; Lance Corporal Dale John LOVERIDGE, Welsh Guards; Sergeant Joseph Gordon MATHER, Special Air Service Regiment; Sergeant Peter Hurdliche Rene NAYA, Royal Army Medical Corps; Warrant Officer Class 2 Brian Thomas NECK, Welsh Guards; Guardsman Andrew Samuel PENGELLY, Scots Guards; Lance Corporal Leslie James Leonard STANDISH, The Parachute Regiment; Sergeant Roman Hugh WREGA, Corps of Royal Engineers.

Mention in Despatches

Sergeant Ian AIRD, The Parachute Regiment; Private Simon John ALEXANDER, The Parachute Regiment; Lieutenant Colonel James ANDERSON, Royal Army Medical Corps; Corporal Raymond Ernest ARMSTRONG, The Royal Green Jackets (Posthumous); Major The Honourable Richard Nicholas BETHEL MBE, Scots Guards; Captain Anthony Peter BOURNE, Royal Regiment of Artillery; Private Andrew Ernest BROOKE, The Parachute Regiment; Driver Mark BROUGH, Royal Corps of Transport; Captain Christopher Charles BROWN, Royal

---

AIR FORCE CROSS
Given the hazards of all military flying, whether or not in the presence of the enemy, this decoration, along with the Air Force Medal, was instituted late in World War I for officers and enlisted personnel respectively for acts of courage and devotion to duty while flying but not actually in combat. They are typically awarded to aircrew who show such qualities in the face of, for example, extreme weather conditions, severe physical exhaustion, or in damaged aircraft. The ribbons reflect those of the DFC and DFM. One DFM was awarded.

ROLL OF HONOUR

Regiment of Artillery; Guardsman Gary BROWN, Scots Guards, Captain Ian Anderson BRYDEN, Scots Guards; Major William Keith BUTLER, Royal Corps of Signals; Staff Sergeant William Henry CARPENTER, Special Air Service Regiment; Lance Corporal Leonard Allan CARVER, The Parachute Regiment; Lieutenant (Queen's Gurkha Officer) CHANDRAKUMAR PRADHAN, 7th Duke of Edinburgh's Own Gurkha Rifles; Staff Sergeant Trevor COLLINS, Corps of Royal Engineers; Private Kevin Patrick CONNERY, The Parachute Regiment; Chaplain to the Forces Third Class David COOPER, Royal Army Chaplains Department; Lieutenant Mark Rudolph CORETH, The Blues and Royals (Royal Horse Guards and 1st Dragoons); Private Adam Michael CORNEILLE, The Parachute Regi-

Regiment (Posthumous); Corporal David FORD, Corps of Royal Engineers; Warrant Officer Class 2 John FRANCIS, Royal Regiment of Artillery; Lieutenant David Peart FRANKLAND, Royal Corps of Transport; Lance Corporal Roy GILLON, Corps of Royal Engineers; Private (now Lance Corporal) Darren John GOUGH, The Parachute Regiment; Lance Sergeant David GRAHAM, Welsh Guards; Private David GRAY, The Parachute Regiment; Major Patrick Hector GULLAN MBE, MC, The Parachute Regiment; Private (Acting Corporal) Joseph Edward HAND, The Parachute Regiment; Lance Corporal (Acting Corporal) Stephen Paul HARDING-DEMPSTER, The Parachute Regiment; Corporal David HARDMAN, The Parachute Regiment (Posthumous); Private Patrick John HARLEY, The Para-

Richard Ryszad KALINSKI, The Parachute Regiment; Captain Simon James KNAPPER, The Staffordshire Regiment (The Prince of Wales's) Staff Sergeant (Acting Warrant Officer Class 2) Anthony LA FRENAIS, Special Air Service Regiment; Major Brendan Charles LAMBE, Royal Regiment of Artillery; Lieutenant Clive Ralph LIVINGSTONE, Corps of Royal Engineers; Lance Corporal Christopher Keith LOVETT, The Parachute Regiment, (Posthumous); Lieutenant Jonathan George Ormsby LOWE, Royal Corps of Transport; Staff Sergeant Clive Dennis LOWTHER, Special Air Service Regiment; Lance Corporal Duncan MACCOLL, Scots Guards; Major Roderick MACDONALD, Corps of Royal Engineers; Piper Peter Alexander MACINNES, Scots Guards; Lance Corporal John Daniel MAHER, Corps of Royal Engineers; Captain Robin John MAKEIG-JONES, Royal Regiment of Artillery; Private Andrew MANSFIELD, The Parachute Regiment; Major Tymothy Alastair MARSH, The Parachute Regiment; Sergeant Peter James MARSHALL, Army Catering Corps; Lance Sergeant Thomas McGUINNESS, Scots Guards; Captain Joseph Hugh McMANNERS, Royal Regiment of Artillery; Lieutenant Alasdair Macfarlane MITCHELL, Scots Guards; Lance Sergeant Clark MITCHELL, Scots Guards (Posthumous); 2nd Lieutenant Ian Charles MOORE, The Parachute Regiment; Private Richard Peter George MORRELL, The Parachute Regiment; Major Philip NEAME, The Parachute Regiment; Corporal Thomas Kiernan NOBLE, The Parachute Regiment; Private Emmanuel O'ROURKE, The Parachute Regiment; Lieutenant Jonathan David PAGE, The Parachute Regiment; Private (Acting Corporal) David John PEARCY, Intelligence Corps; Corporal Jeremy Frank PHILLIPS, The Parachute Regiment; Private (Acting Sergeant) Brian William PITCHFORTH, The Queen's Regiment; Private Anthony POTTER, Royal Army Ordnance Corps; Lance Corporal Barry John RANDALL, Corps of Royal Engineers; Sergeant Peter RATCLIFFE, Special Air Service Regiment; Lance Corporal Graham RENNIE, Scots Guards; Warrant Officer Class 2 Malcolm Douglas RICHARDS, Royal Regiment of Artillery; Lance Corporal Julian Jon RIGG, Army Air Corps; Lieutenant Colonel John David Arthur ROBERTS, Royal Army Medical Corps; Major Barnaby Peter Stuart ROLFE-SMITH, The Parachute Regiment; Captain Christopher Roy ROMBERG, Royal Regiment of Artillery; Lieutenant Colonel Hugh Michael ROSE, OBE, Coldstream Guards; Sergeant Ian ROY, Corps of Royal Engineers; Captain Julian David Gurney SAYERS, Welsh Guards; Lieutenant (Acting Captain) Matthew Rodgers SELFRIDGE, The Parachute Regiment (Posthumous); Warrant Officer Class 2 Michael John SHARP, Army Air Corps; Corporal John William SIBLEY, The Parachute Regiment; Major Colin Stewart SIBUN, Army Air Corps; Sapper (Acting Lance Corporal) William Austen SKINNER, Corps of Royal Engineers; Major Graham Frederick William SMITH, Royal Regiment of Artillery; Captain Royston John SOUTHWORTH, Royal Army Ord-

nance Corps; Corporal of Horse Paul STRETTON, The Blues and Royals (Royal Horse Guards and 1st Dragoons); 2nd Lieutenant James Douglas STUART, Scots Guards; Lieutenant William John SYMS, Welsh Guards; Corporal (Acting Sergeant) Robert Clive TAYLOR, Royal Corps of Signals; Major Anthony TODD, Royal Corps of Transport; Lance Corporal Gary TYTLER, Scots Guards; Private (Acting Corporal) Peter Andrew WALKER, The Staffordshire Regiment (The Prince of Wales's); Sergeant Richard John WALKER, Army Air Corps; 2nd Lieutenant Guy WALLIS, The Parachute Regiment; Lieutenant Mark Evan WARING, Royal Regiment of Artillery; Captain James Nicholas Edward WATSON, Royal Regiment of Artillery; Lieutenant Geoffrey Ronald WEIGHELL, The Parachute Regiment; Lieutenant (now Captain) Mark Graham WILLIAMS, Royal Regiment of Artillery; Lieutenant (now Captain) Maldwyn Stephen Henry WORSLEY-TONKS, The Parachute Regiment.

ROYAL AIR FORCE

Distinguished Service Cross
Flight Lieutenant David Henry Spencer MORGAN, Royal Air Force, 899 Naval Air Squadron.

Distinguished Flying Cross
Wing Commander Peter Ted SQUIRE AFC, Royal Air Force; Squadron Leader Richard Ulric LANGWORTHY AFC, Royal Air Force; Squadron Leader Calum Neil McDOUGALL, Royal Air Force; Squadron Leader Jeremy John POOK, Royal Air Force; Flight Lieutenant William Francis Martin WITHERS, Royal Air Force.

From the citation for
Wing Commander
P T Squire's DFC

'Wing Commander Squire led his Squadron with great courage from the front, flying 24 attack sorties. He flew many daring missions, but of particular note was an attack at low level with rockets on targets at Port Stanley Airfield in the face of heavy anti-aircraft fire when both he and his wing man returned damaged. Also a bombing attack on an HQ position when, on approach, a bullet passed through his cockpit which temporarily distracted him, but he quickly found an alternative target and bombed that instead'.

Air Force Cross
Wing Commander David EMMERSON, Royal Air Force; Squadron Leader Arthur Max ROBERTS, Royal Air Force; Squadron Leader Robert TUXFORD, Royal Air Force; Flight Lieutenant Harold Currie BURGOYNE, Royal Air Force.

Queen's Gallantry Medal
Flight Lieutenant Alan James SWAN, Royal Air Force; Flight Sergeant Brian William JOPLING, Royal Air Force.

MENTION IN DISPATCHES

QUEEN'S COMMENDATION FOR BRAVE CONDUCT

QUEEN'S COMMENDATION FOR VALUABLE SERVICE IN THE AIR

Servicemen of all ranks who distinguish themselves by some act of courage or skill which is not held to qualify them for the

award of a higher honour receive these commendations. Like all other decorations they are awarded on the recommendation of unit commanders—though once again, like all other decorations, far fewer are awarded than are recommended. The visible mark of this class of honour is an oak leaf insignia, which is worn by service personnel sewn to the ribbon of the campaign medal relevant to the award, or directly to the tunic if no campaign is involved.

Although most of these medals and decorations are specific to one or other of the fighting services, officers and men of other services may receive them if their duties qualify them. For instance, in these days of combined operations, a Royal Marine or Army helicopter pilot might be awarded a DFC or DFM; and naval or air personnel serving on the ground as specialist or liaison personnel attached to an Army unit might be awarded Army decorations for their conduct during ground combat.

---

ment; Corporal Ian Clifford CORRIGAN, Corps of Royal Electrical and Mechanical Engineers; Lieutenant Mark Townsend COX, The Parachute Regiment; Staff Sergeant Phillip Preston CURRASS, QGM, Royal Army Medical Corps (Posthumous); Lance Sergeant Alan Charles Ewing DALGLEISH, Scots Guards; Lance Corporal Neal John DANCE, The Parachute Regiment; Lance Sergeant Ian DAVIDSON, Scots Guards; Major Peter Eastaway DENNISON, The Parachute Regiment; Staff Sergeant George Kenneth DIXON, Royal Regiment of Artillery; Piper Steven William DUFFY, Scots Guards; Lance Corporal Kevin Peter DUNBAR, The Parachute Regiment; Gunner Gary ECCLESTON, Royal Regiment of Artillery; Captain Martin Patrick ENTWISTLE, Royal Army Medical Corps; Lieutenant Colonel Keith Richard Hubert EVE, Royal Regiment of Artillery; Captain Paul Raymond FARRAR, The Parachute Regiment; Private Mark William FLETCHER, The Parachute

chute Regiment; Major Richard Bruce HAWKEN, Corps of Royal Engineers; Lieutenant Robert Charles HENDICOTT, Corps of Royal Engineers; Corporal (Acting Sergeant) Joseph HILL, The Parachute Regiment; Lieutenant Colonel George Anthony HOLT, Royal Regiment of Artillery; Warrant Officer Class 2 Graham HOUGH, Welsh Guards; Captain (now Major) Euan Henry HOUSTOUN MBE, Grenadier Guards; Lance Bombardier (Acting Bombardier) Owain Dyfnallt HUGHES, Royal Regiment of Artillery; Captain Stephen James HUGHES, Royal Army Medical Corps; Corporal Stephen Darryl ILES, Corps of royal Engineers; Lieutenant The Lord Robert Anthony INNESKER, The Blues and Royals (Royal Horse Guards and 1st Dragoons); Bombardier John Rodney JACKSON; Royal Regiment of Artillery; Gunner Jeffrey JONES, Royal Regiment of Artillery; Lance Corporal Kenneth Bryan JONES, Royal Corps of Transport; Sergeant

Flt Lt Alan Swann, QGM

Queen's Commendation for Valuable Service in the Air

Squadron Leader Timothy Newell ALLEN, Royal Air Force; Squadron Leader Anthony Frank BANFIELD, Royal Air Force; Squadron Leader Geoffrey Roger BARRELL, Royal Air force; Flight Lieutenant John Allin BROWN, Royal Air Force; Flight Lieutenant Peter Alfred STANDING, Royal Air Force; Squadron Leader (now Wing Commander) Martin Donald TODD, Royal Air Force; Squadron Leader Ernest Frederick WALLIS, MBE Royal Air Force; Flight Lieutenant Michael Ernest BEER, Royal Air Force; Flight Lieutenant James Dalrymple CUNNINGHAM, Royal Air Force; Flight Lieutenant John Norman KEABLE, Royal Air Force; Flight Lieutenant Murdo MacDonald MacLEOD, Royal Air Force; Flight Lieutenant Glyn David REES, Royal Air Force; Flight Lieutenant Robert Leslie ROWLEY, Royal Air Force; Flight Sergeant Stephen Edward SLOAN, Royal Air Force.

Mention in Despatches

Squadron Leader John Geoffrey ELLIOTT, Royal Air Force; Squadron Leader Robert Douglas IVESON, Royal Air Force; Flight Lieutenant Edward Henry BALL, Royal Air Force; Flight Lieutenant Mark William James HARE, Royal Air Force; Flight Lieutenant Gordon Carnie GRAHAM, Royal Air Force; Flight Lieutenant Alan Tom JONES, Royal Air Force; Flight Lieutenant Ian MORTIMER, Royal Air Force; Flight Lieutenant Hugh PRIOR, Royal Air Force; Flight Lieutenant Richard John RUSSELL, AFC Royal Air Force; Flight Lieutenant Robert Dennis WRIGHT, Royal Air Force; Flying Officer Peter Lewis TAYLOR, Royal Air Force; Flying Officer Colin MILLER, Royal Air Force; Flight Sergeant Derek William KNIGHTS, Royal Air Force; Corporal Alan David TOMLINSON, Royal Air Force

Queens's Commendation for Brave Conduct

Junior Technician Adrian THORNE, Royal Air Force; Senior Aircraft man Kenneth James SOPPETT-MOSS, Royal Air Force.

CIVILIAN
Life Peer
Admiral of the Fleet Sir Terence Thornton LEWIN GCB, MVO, DSC, lately Chief of the Defence Staff.

Knight Batchelor
Rex Masterman HUNT CMG, HM Civil Commissioner, Falkland Islands.

CB (Civil Division)
Kenneth John PRITCHARD, Assistant Under Secretary, Ministry of Defence.

CMG
David Heywood ANDERSON, Foreign and Commonwealth Office.

Order of the Bath (Military Division)
KCB
Major General John Jeremy MOORE, CB, OBE, MC; Rear Admiral John Forster WOODWARD.

CB
Air Vice-Marshal George Arthur CHESWORTH, OBE, DFC, Royal Air Force; Commodore Michael Cecil CLAPP, Royal Navy; Air Vice-Marshal Kenneth William HAYR, CBE, AFC, Royal Air Force; Brigadier Julian Howard Atherden THOMPSON, OBE, ADC, Royal Marines; Rear Admiral Anthony John WHETSTONE;

Order of the British Empire (Military Division) GBE
Admiral Sir John David Elliott FIELDHOUSE, GCB

KBE
Air Marshal Sir John Bagot CURTISS, KCB, Royal Air Force; Vice Admiral David John HALLIFAX

CBE
Captain Paul BADCOCK, Royal Navy; Captain Nicholas John BARKER, Royal Navy; Colonel Ian Stuart BAXTER, MBE, late Royal Corps of Transport; Colonel John David BIDMEAD, OBE, late Royal Corps of Transport; Captain Christopher Peter Oldbury BURNE, Royal Navy; Colonel (Now Brigadier) David Bryan Hall COLLEY OBE, late Royal Corps of Transport; Group Captain Clive Ernest EVANS, Royal Air Force; Captain Raymnd Hunter FOX, Royal Navy; Captain John GARNIER, MVO, Royal Navy; Group Captain Alexander Freeland Cairns HUNTER, OBE, AFC, Royal Air Force; Group Captain Patrick KING, OBE, Royal Air Force; Captain Michael Henry Gordon LAYARD, Royal Navy; Colonel Bruce Christopher McDERMOTT, OBE, late Royal Army Medical Corps; Captain Robert McQUEEN, Royal Navy; Group Captain Jeremy Simon Blake PRICE, ADC, Royal Air Force; Captain Jonathan James Richard TOD, Royal Navy; Captain John Peter WRIGLEY, Royal Navy.

OBE
Commander Thomas Anthony ALLEN, Royal Navy; Wing Commander Anthony John Crowther BAGNALL, Royal Air Force; Commander Lional Stuart Joseph BARRY, Royal Navy; Wing Commander David Llewellyn BAUGH, Royal Air Force; Lieutenant Colonel Anthony Edward BERRY, The Royal Green Jackets; Commander Peter Stanley BIRCH, Royal Navy; Major Robert James BRUCE, Royal Marines; Major John Shane CHESTER, Royal Marines; Commander Michael CUDMORE, Royal Navy; Captain John Barrie DICKINSON, Royal Fleet Auxiliary; Wing Commander Peter FRY, MBE, Royal Air Force; Commander Frederick Brian GOODSON, Royal Navy; Lieutenant Colonel Ivar Jack HELLBERG, Royal Corps of Transport; Commander Lister Theodore HICKSON, Royal Navy; Major (Now Lieutenant Colonel) Peter John HUBERT MBE, The Queen's Regiment; Surgeon Lieutenant Commander (Acting Surgeon Commander) Richard Tadeusz JOLLY, Royal Navy; Commander Christopher John ESPLIN-JONES, Royal Navy; Captain John Stuart KELLY, MBE, Royal Navy; Commander David Arthur Henry KERR, Royal Navy; Commander Martin Leonard LADD, Royal Navy; Captain Peter James McCARTHY, Royal Fleet Auxiliary; Commander Peter John McGREGOR, Royal Navy; Lieutenant Colonel William Stewart Petrie McGREGOR, Royal Army Medical Corps; Major David John MINORDS, Royal Marines; Lieutenant Colonel David Patrick de Courcy MORGAN, 7th Duke of Edinburgh's Own Gurkha Rifles; Squadron Leader Brian Sydney MORRIS, AFC, Royal Air Force; Commander Andrew William NETHERCLIFT, Royal Navy; Commander (Acting Captain) Anthony James OGLESBY, Royal Navy; Captain Gilbert Paul OVERBURY, Royal Fleet Auxiliary; Commander George Sheddon PEARSON, Royal Navy; Captain Shane REDMOND, Royal Fleet Auxiliary, Lieutenant Colonel John Francis RICKETT, MBE, Welsh Guards; Commander Andrew Stephen RITCHIE, Royal Navy; The Reverend Anthony McPherson ROSS, Royal Navy; Commander Robert Austin ROWLEY, Royal Navy; Commander Jeremy Thomas SANDERS, Royal Navy; Commander Ronald James SANDFORD, Royal Navy; Lieutenant Colonel (Quartermaster) Patrick John SAUNDERS, Corps of Royal Engineers; Major James Maurice Guy SHERIDAN, Royal Marines; Commander Donald William SHRUBB, Royal Navy; Wing Commander Joseph KERR, AFC, Royal Air Force; Wing Commander Anthony Peter SLINGER, Royal Air Force; Lieutenant Colonel Michael John HOLROYD SMITH, Royal Regiment of Artillery; Wing Commander Charles Julian STURT, Royal Air Force; Major Simon Ewen SOUTHBY-TAILYOUR, Royal Marines; Major Jonathan James THOMSON, Royal Marines; Wing Commander Brian James WEAVER, Royal Air Force; Lieutenant Colonel Ronald WELSH, Royal Army Medical Corps; Commander Christopher Watkin WILLIAMS, Royal Navy; Commander George Anthony Charles WOODS, Royal Navy; Captain Christopher Anthony PURTCHER-WYDENBRUCK, Royal Fleet Auxiliary.

Warrant Officer Daniel Philmore BARKER, Royal Air Force; Major Edward Leo BARRETT, Royal Corps of Transport; Major Charles Gordon BATTY, Royal Army Medical Corps; Lieutenant Commander Michael John Douglas BROUGHAM, Royal Navy; Lieutenant Commander Roger Charles CAESLEY, Royal Navy; Flight Lieutenant Edna May CLINTON, Women's Royal Air Force; Lieutenant Roger Stephen COLLINS, Royal Navy; Warrant Officer Class 1 (Regimental Sergeant Major) Anthony James DAVIES, Welsh Guards; Major Christopher Matthew DAVIES, Corps of Royal Engineers; Lieutenant Alan David DUMMER, Royal Navy; Flight Lieu-

REVERSE

REVERSE

GEORGE MEDAL
QUEEN'S GALLANTRY MEDAL

These two medals, together with the George Cross (not in fact awarded in the Falklands War), are outside the normal sequence of awards for behaviour in combat on land, sea and air.

The GM is awarded to both civilian and service personnel for acts of exceptional bravery in circumstances not covered by the award of strictly military honours. It may, for example, be given for extraordinary bravery in bomb disposal work. The GC can be described as a civilian VC.

Instituted in 1974, the Queen's Gallantry Medal replaces the previous award of the British Empire Medal for gallantry. It is primarily a civilian honour, but, like the GC and GM, can be awarded to service personnel for acts which do not meet the criteria for specifically military decorations. It is listed after the DFM in order of precedence, and is often awarded in peacetime, as well as for suitable acts in war but not in the course of active operations.

tenant John DUNGATE, AFM, Royal
Air Force; Major John Anthony
EAST, Royal Army Medical Corps;
Lieutenant Commander Colin John
EDWARDS, Royal Navy; Fleet Chief
Radio Supervisor David John
EGGERS, Warrant Officer Class 1
Leslie ELLSON, Welsh Guards;
Lieutenant Simon Jonathan BRANCH-
EVANS, Royal Navy; Major Andrew
Roger GALE, Royal Corps of Signals;
Lieutenant Commander Richard
GOODENOUGH, Royal Navy; Lieu-
tenant Commander Michael
GOODMAN, Royal Navy; Major
Charles GRIFFITHS, Royal Army
Medical Corps; Major (Quarter-
master) Gerald Norman GROOM,
Royal Corps of Transport; Warrant
Officer Class 2 (Acting Warrant
Officer Class 1) Thomas HAIG,
Special Air Service Regiment; Lieu-
tenant Commander Robert William
HAMILTON, Royal Navy; Major
Laurence HOLLINGWORTH, Royal
Army Ordnance Corps; Captain Colin
Francis HOWARD, Royal Marines;
Lieutenant Commander Gerard
Martin John IRVINE, Royal Navy;
Lieutenant Commander Peter John
JAMES, Royal Naval Reserve;
Squadron Leader Clive Graham
JEFFORD, Royal Air Force; Fleet
ChiefWriter Christopher Geoffrey
LAMB; Fleet Chief Petty Officer (OPs)
(S) Michael John LEGG; Squadron
Leader William Frederick LLOYD,
Royal Air Force; Lieutenant Com-
mander James Hutcheon LOUDON,
Royal Navy; Captain Ronald
MARSHALL, Intelligence Corps;
Flight Lieutenant Brian Thomas
MASON, Royal Air Force; Lieutenant
Commander Horace Alfred MAYERS,
Royal Navy; Captain Terence Gerald
McCABE, Royal Army Medical
Corps; Warrant Officer Class 1
Michael John McHALE, Royal Army
Medical Corps; Lieutenant Com-
mander Ian Scott McKENZIE, Royal
Navy; Lieutenant Commander James
Murdoch MILNE, Royal Navy; Cap-
tain (Quartermaster) Norman
Edward MENZIES, The Parachute
Regiment; Lieutenant (Now Captain)
Frederick James MOODY, Scots
Guards; Warrant Officer Class 2
Derek MOORE, Royal Corps of
Transport; Fleet Chief Marine En-
gineering Artificer (H) Peter William
MULLER; Acting Flight Lieutenant
Anthony NEALE, Royal Air Force;
Fleet Chief Petty Officer (OPS) (S)
Robert John NICHOLLS; Squadron
Leader David Miller NIVEN, Royal
Air Force; Lieutenant David Charles
Winston O'CONNELL, Royal Navy;
Lieutenant Commander Lawrence
David POOLE, Royal Navy; Lieu-
tenant Brian PURNELL, Royal Navy;
Warrant Officer Class 1 Robin Glen
RANDALL, Royal Corps of Engi-
neers; Major (Ordnance Executive
Officer) John Moorby RIDDING,
Royal Army Ordnance Corps; Flight
Lieutenant Paul Anthony ROOM,
Royal Air Force; Captain Michael
Jeremy SHARLAND, Royal Marines;
Surgeon Lieutenant Commander
Philip James SHOULER, Royal Navy;
Squadron Leader Trevor SITCH,
Royal Air Force; Master Air Load-
master Alan David SMITH, Royal Air
Force; Lieutenant David Fielding
SMITH, Royal Navy; Lieutenant
Commander David John Robert
WILMOT-SMITH, Royal Navy; Cap-
tain Dennis SPARKS, Royal Marines;
Squadron Leader John Edward
STOKES, Royal Air Force; Major
John Ronaldson STUART, Royal

ORDER OF THE BATH
ORDER OF THE BRITISH EMPIRE
ORDER OF ST MICHAEL AND ST GEORGE

1 CB *(military)*; 2 OBE *(military)*;
3 CMG; 4 MBE *(civilian)*; 5 CBE
(military)
These awards complete the
Falklands honours picture. The
first two are awarded both to
military personnel and civilians.

The Most Honourable Order
of the Bath, tracing a tenuous
ancestry back to a court
appointment instituted in 1399,
was revived in 1725 as an award
for senior officers for service in
action. There are three classes:

Knight Grand Cross (GCB);
Knight (GCB) and Companion
(CB). The senior two classes
carry knighthoods as the titles
imply. The KCB has traditionally
been the reward for the
commander-in-chief of
victorious British forces. The CB
can only be conferred on an
officer of the equivalent rank of
major, commander or higher.

The Most Excellent
Order of the British Empire was
instituted in 1917, and a military
division was created the
following year. It is primarily a
civilian award, conferred for a
wide range of services to the
nation, in five grades: GBE, KBE
(both grades carrying
knighthoods), CBE, OBE, and
MBE. It is conferred in the
various grades on military
personnel of all ranks for
valuable service which does not
qualify, by reason of
circumstance, for any
specifically military decoration.
In honours lists such awards are
usually described as being in the
'military division'.

The Most Distinguished
Order of St Michael and St
George was instituted in 1818
and is awarded, in four classes,
traditionally but not exclusively,
for overseas service. As such it is
the principal award for
diplomats.

Corps of Signals; Major Michael
Gordon TAYLOR, Royal Corps of
Signals; Lieutenant Commander
John Nicholas Owen WILLIAMS,
Royal Navy; Warrant Officer Class 2
Philip Michael WILLIAMS, Royal
Corps of Transport; Major Timothy
James WILTON, Royal Regiment of
Artillery; Major Guy Justin
YEOMAN, Royal Corps of Transport;
Warrant Officer Class 2 Robert
Charles YEOMANS, Royal Corps of
Signals.

British Empire Medal (Military Division)
Petty Officer Medical Assistant Keith
ADAMS; Air Engineering Mechani-
cian (R) 1st Class John Leslie BAILEY;
Chief Air Engineering Mechanic (M)
Norman Ronald BARWICK; Flight
Sergeant John Harry BELL, Royal Air
Force; Marine Engineering Artificer
(H) 1st Class Thomas James
BENNETTO; Staff Sergeant William
Frank BLYTH, Royal Corps of
Transport; Staff Sergeant Edward
George BRADBURY, Corps of Royal
Engineers; Sergeant Roger Joseph
BROWN, Corps of Royal Engineers;
Chief Air Engineering Artificer (R) 1st
Class David Martine CHILDS;
Master-at-Arms Anthony Francis
COLES; Sergeant James McMillan
COLEMAN, Royal Air Force; Chief
Marine Engineering Mechanician (P)
Geoffrey Stuart COX; Staff Sergeant
Michael John DENT, Corps of Royal
Engineers; Chief Air Engineering
Mechanic (L) William David EATON;
Staff Sergeant James FENWICK,
Corps of Royal Electrical and Mecha-
nical Engineers; Air Engineering
Artificer (M) 1st Class Stuart John
GOODALL; Staff Sergeant Robert
Leonard GRIFFITHS, Royal Corps of
Signals; Corporal Norman John
HALL, Corps of Royal Engineers;
Sergeant David HARVEY, Royal
Army Ordnance Corps; Staff
Sergeant Colin Lee HENDERSON,
Army Catering Corps; Chief Air
Engineering Aritificer (M) David
John HERITIER; Corporal Graham
John HERRINGTON, Royal Pioneer
Corps; Chief Petty Officer (D)
Leonard Brian HEWETT; Staff Ser-
geant John Duncan HOLMES, Royal
Army Ordnance Corps; Corporal
William Henry HOPKINS, Royal
Army Ordnance Corps; Private David
John HUNT, Army Catering Corps;
Chief Petty Officer Caterer John
Arthur JACKSON; Air Engineering
Artificer (H) 1st Class David Eric
JONES; Flight Sergeant Kenneth
KENNY, Royal Air Force; Chief
Technician Thomas Joseph
KINSELLA, Royal Air Force; Air
Engineering Artificer (L) 1st Class
Robert Anthony John MASON;
Medical Technician 1st Class Stuart
McKINLAY; Chief Petty Officer Cook
Michael Gerald MERCER; Leading
Wren Stores Accountant Jacqueline
MITTON, Women's Royal Naval
Service; Chief Wren Education
Assistant Anne MONCTON, Women's
Royal Naval Service; Sergeant Denis
Ronald PASFIELD, Corps of Royal
Engineers; Staff Sergeant Paul
RAYNER, Corps of Royal Engineers;
Staff Sergeant (Acting Warrant
Officer Class 2) Malachi REID, Royal
Army Medical Corps; Air Engineer-
ing Artificer (L) 2nd Class Alan John
SMITH; Chief Petty Officer (OPS) (M)
Owen Gwyn STOCKHAM; Air
Engineering Aritificer (L) 1st Class
Roger James Edward STRONG; Staff
Sergeant Christopher Glyn TAYLOR,

Corps of Royal Electrical and Mechanical Engineers; Chief Air Engineering Mechanician (L) 1st Class Thomas Lowen TEMPLE; Leading Wren Dental Hygienist Kim TOMS, Women's Royal Naval Service; Chief Wren Family Services Barbara Marion TRAVERS, Women's Royal Naval Service; Sergeant Peter TUXFORD, Royal Air Force; Chief Technician Richard Keith VERNON, Royal Air Force; Sergeant John Charles VICKERS, Royal Air Force; Corporal David John VIVIAN, Royal Air Force; Acting Leading Stores Accountant Gerard John WALSH; Petty Officer (Missile) John James Trevor WATERFIELD; Petty Officer (Missile) Edward Lee WELLS; Air Engineering Mechanician (M) 1st Class David John WILLIAMS; Sergeant Brian WINTER, Royal Marines; Corporal (Acting Sergeant) Anthony WORTHINGTON, Corps of Royal Engineers.

CBE (Civil Division

Captain Donald Arthur ELLERBY, Master m.v. NORLAND; Ian McLeod FAIRFIELD, Chairman and Chief Executive, Chemring plc; Miss Patricia Margaret HUTCHINSON, CMG HM Ambassador, Montevideo; Roger Tustin JACKLING, Assistant Secretary, Ministry of Defence; Captain Dennis John SCOTT-MASSON, Master SS CANBERRA; Captain John Penny MORTON, Master, m.v. ELK; Nigel Hamilton NICHOLLS, Assistant Secretary, Ministry of Defence; Eric John RISNESS, Deputy Chief Scientific Officer, Ministry of Defence; William Bell SLATER, Managing Director, The Cunard Steam-Ship Company plc; John Robert Christopher THOMAS, Deputy Chief Scientific Officer, Ministry of Defence.

OBE

Peter Derek ADAMS, Principal Scientific Officer, Ministry of Defence; Russell George ALGAR, Senior Principal, Ministry of Defence; The Reverend Harry BAGNALL, Dean of Christchurch, Falkland Islands; Michael John BEYNON, Chief Map Research Officer, Ministry of Defence; Alison Ann, Mrs BLEANEY, Acting Senior Medical Officer, Falkland Islands; Margaret Janet, Mrs BOURNE, Senior Principal Scientific Officer, Ministry of Defence; Reginald BUTCHER, Managing Director, Wimpey Marine Ltd; David William CHALMERS, Constructor (C), Ministry of Defence; Captain William James Christopher CLARKE, Master, m.v. EUROPIC FERRY; Captain Alan FULTON, Master, Cable Ship IRIS; Roderick Owen GATES, Executive Director, Aircraft Engineering, Marshall of Cambridge (Engineering) Ltd; Andrew John GLASGOW, Projects Director, Marconi Underwater Systems Ltd; Edgar James HARVEY, Principal Professional and Technology Officer, Ministry of Defence; Stanley Stephen HOLNESS, Senior Principal, Department of Trade; Vernon Edward HORSFIELD, Works Manager, Woodford Aircraft Group, British Aerospace plc; Christopher HULSE, Foreign and Commonwealth Office; Miss Maureen Mary JONES, Foreign and Commonwealth Office; Derek LEWIS, Professional and Technology Superintendent, Ministry of Defence; Arthur Frederick George MOSS, Divisional Manager,

HM Dockyard, Gibralter; John Patrick RABY, Projects Director, Humber Graving Dock and Engineering Company Ltd; Captain David Michael RUNDLE, Master, m.v. BRITISH WYE; Captain Michael John SLACK, Master, m.s. WIMPEY SEAHORSE; The Right Reverend Monsignor Daniel Martin SPRAGGON, MBE, Prefect Apostolic, Falkland Islands; Raymond Sydney TEE, Principal Professional and Technology Officer (Constructor), Ministry of Defence; Peter VARNISH, Principal Scientific Officer, Ministry of Defence; Ronale WATSON, Local Director, Quality Assurance, Swan Hunter Shipbuilders Ltd; Robert WEATHERBURN, Senior Principal Scientific Officer, Ministry of Defence; John Anthony WELDON, Principal Professional and Technology Officer, Ministry of Defence.

SOUTH ATLANTIC MEDAL
The ribbon of the campaign medal, awarded to all who took part, bears a rosette if the recipient entered the combat zone.

MBE

Valerie Elizabeth, Mrs BENNETT, Acting Matron, Stanley Hospital, Falkland Islands; Jane Hunter, Mrs BOLTON, Clerical Officer, Ministry of Defence; Colin Michael BOYNE, Senior Scientific Officer, Ministry of Defence; David Laing BREEN, Radar Systems Engineer, Marconi Radar Systems Ltd; Ronald Arthur BROWN, Marine Services Officer II (Engineer), Ministry of Defence; Terence James CAREY, Electrical Superintendent, Falkland Islands; Edgar Dennis CARR, Regional Manager, Southampton, General Council of British Shipping; Anthony Martin CLEAVER, Photographer, Press Association; Albert Frederick George COLLINS, Steelwork Production Manager, Vosper Ship Repairers Ltd; Arthur John COLLMAN, Professional and Technology Officer II, Ministry of Defence; Peter Merlyn John COOK, Professional and Technology Officer II, Ministry of Defence; Frederick Joseph COOPER, Passenger and Cargo Manager's Assistant, British Transport Docks Board, Southampton; David John CORMICK, Senior Field Engineer, Marconi Space and Defence Systems Ltd; Richard Arthur DREW, Foreign and Commonwealth Office; Miss Patricia DURLING, Higher Executive Officer, Ministry of Defence; Stuart EARNSHAW, Chief Marine Superintendent, Thoresen Car Ferries Ltd; Miss Mary Georgiana ELPHINSTONE, Volunteer Medical Officer, Falkland Islands; Miss Rosemary Margaret ELSDON, Senior Nursing Sister, SS CANBERRA; James Robert Rutherford FOX, Radio News Reporter, British Broadcasting Corporation; John Aubrey FRENCH, Senior Scientific Officer, Ministry of

Defence; Brian Arthur GORRINGE, Catering Manager Grade II, Staff Restaurant, Ministry of Defence; Eric Miller GOSS, Manager, Goose Green Farm, Falkland Islands; Michael John Stephen HATTON, Professional and Technology Officer II, Department of Trade; Miss Sybil Matilda HILL, Clerical Assistant, Department of Trade; Gerald William Tom HODGE, Professional and Technology Officer II, Ministry of Defence; William HUNTER, Professional and Technology Officer II, Ministry of Defence; Ronald Daniel LAWRENCE, Higher Executive Officer, Cabinet Office; Robert Graham John LLOYD, Assistant Manager, Warehouse and Distribution Services, Navy, Army and Air Force Institutes; David McALPIN, Flight Trials Engineer, Ferranti plc; William Robert McQUEEN, Senior Scientific Officer, Meteorological Office; David MONUMENT, Maintenance Superintendent, P&O Steam Navigation Company; Thomas Ronald MORSE, Foreign and Commonwealth Office; Valerie Ann, Mrs MOTHERSHAW, Executive Officer, Ministry of Defence; Dawn Barbara Mavis, Mrs MURRAY, Senior Scientific Officer, Ministry of Defence; Patricia Margaret, Mrs NUTBEEM, Chairwoman, 16 Field Ambulance RAMC Wives' Club, Aldershot; Squadron Leader Thomas James PALMER, RAF (Retd), Headquarters, United Kingdom Land Forces, Ministry of Defence; Miss Elizabeth PATTEN, Senior Welfare Officer, St John and Red Cross Service Hospitals Welfare; Terence John PECK CPM, Councillor, Legislative Council, Falkland Islands; Denis PLACE, Water Supervisor, Falkland Islands; Jonathan Trevor PRICE, Executive Officer, Ministry of Defence; James Frederick QUIRK, Senior Executive Officer, Royal Naval Supply and Transport Service; Paul ROBINSON, Higher Scientific Officer, Ministry of Defence; John Robertson Page RODIGAN, Professional and Technology Officer II, Ministry of Defence; Kevin William SHACKLETON, Contract Engineer, Ames Crosta Babcock Ltd; Michael Sydney SHEARS, Production Manager, Vosper Thornycroft (UK) Ltd; Captain Derek SIMS, Senior Cargo Surveyor, Hogg Robinson (GFA) Ltd; Miss Angela SLAYMAKER, Clerical Officer, Ministry of Defence; Squadron Leader John Michael SMITH, RAF (Retd), Senior Operations Manager, Dynamics Group, British Aerospace plc; Rodney Lorraine START, Senior Executive Officer, Department of Trade; Angela Elizabeth, Mrs THORNE, Executive Officer, Ministry of Defence; John TURNER, Senior Scientific Officer, Meteorological Office; Patrick James WATTS, Director, Broadcasting Service, Falkland Islands; Richard Stephen WHITLEY, Veterinary Officer, Falkland Islands.

British Empire Medal (Civil Division)

Arthur James ALDRED, Process and General Supervisory Grade D, Ministry of Defence; Malcolm ASHWORTH, Dairyman, Falkland Islands; Garry BALES, Able Seaman, Tug IRISHMAN; Irene Ingeborg, Mrs BARDSLEY, Club Manageress, Excellent Steps, Portsmouth, Navy, Army and Air Force Institutes; Richard Sydney BARRETT, Chief Steward,

Cable Ship IRIS; Dennis Paul BETTS, Able Seaman, Tug IRISHMAN; Roy Samuel BLANCHARD, Foreman Shipwright, Vosper Ship Repairers Ltd; Michael Harfield BOYES, Laboratory Mechanic, Ministry of Defence; Najla Dorothy, Mrs BUCKETT, Housewife, Falkland Islands; Tim DOBBYNS, Farmer, Falkland Islands; Eric Christopher EMERY, Professional and Technology Officer III, Department of Trade; Luis ESTELLA, Process and General Supervisory Grade E, HM Dockyard, Gibralter; James Stephen FAIRFIELD, lately Corporal, Royal Marines, Falkland Islands; Robert James FORD, Senior Storeman, Ministry of Defence; James Anderson GOLDIE, Stores Officer Grade C, Royal Fleet Auxiliary RESOURCE; Leslie Sydney HARRIS, Senior Electrician, Falkland Islands; Ronald John HATCH, Marine Services Officer IV (Deck), Ministry of Defence; John HAYWOOD, Progressman Planner Technical (Shipwright), Ministry of Defence; Jack JOHNSTON, Senior Storekeeper, Royal Fleet Auxiliary FORT AUSTIN; James Frederick JONES, Professional and Technology Officer III, Ministry of Defence; Bernard ORAM-JONES, Shipwright, Ministry of Defence; Brian John JOSHUA, Catering Manager, Pan American Airways, United States Air Force Base, Ascension Island; KANG, SHIK-MING, Laundryman, HMS BRILLIANT; Gordon James LANE, Laboratory Mechanic, Ministry of Defence; Alan John LEONARD, Chief Cook, SS ATLANTIC CAUSEWAY; Joseph Anthony LYNCH, Stores Officer Grade C, Ministry of Defence; Paul McEWAN, Stores Officer Grade C, Royal Fleet Auxiliary REGENT; Michael McKAY, Farmer, Falkland Islands; Philip MILLER, Tractor Driver, Falkland Islands; Edwin George MORGAN, Professional and Technology Officer III, Ministry of Defence; Andrew James Graham NISBET, Professional and Technology Officer III, Ministry of Defence; Hilda Blanche, Mrs PERRY, Telephone Superintendent, Falkland Islands; Peter Richard PETERSON, Mechanical Fitter, David Brown Gear Industries; Raymond Arthur ROBJOHN, Superintendent, Experimental Flight Shed, Westland Helicopters; Derek Robert Thomas ROZEE, Plumber, Falkland Islands; Ellis Walton SAMPSON, Stores Officer Grade C, Ministry of Defence; Victor SEOGALUTZE, Assistant Chief Inspector, Bridport Gundry plc; David Albert SMERDON, Professional and Technology Officer IV, Ministry of Defence; Vernon STEEN, Guide, Falkland Islands; SUEN, Ling-Kan, Laundryman, HMS ANTRIM; Donald Victor THREADGOLD, Telecommunications Technical Officer Grade II, Ministry of Defence; Miss Karen Lois TIMBERLAKE, Nursing Sister, Falkland Islands; Roland TODD, Professional and Technology Officer III, Ministry of Defence; Frank John TOUGH, Professional and Technology Officer III, Ministry of Defence; Eileen, Mrs VIDAL, Radio Telephonist, Falkland Islands; Miss Bronwen Vaughan WILLIAMS, Nursing Sister, Falkland Islands; Colin Walter WILSON, Foreman, Repair Support Area, Marconi Radar Systems; Christopher John WINDER, Professional and Technology Officer III, Ministry of Defence.

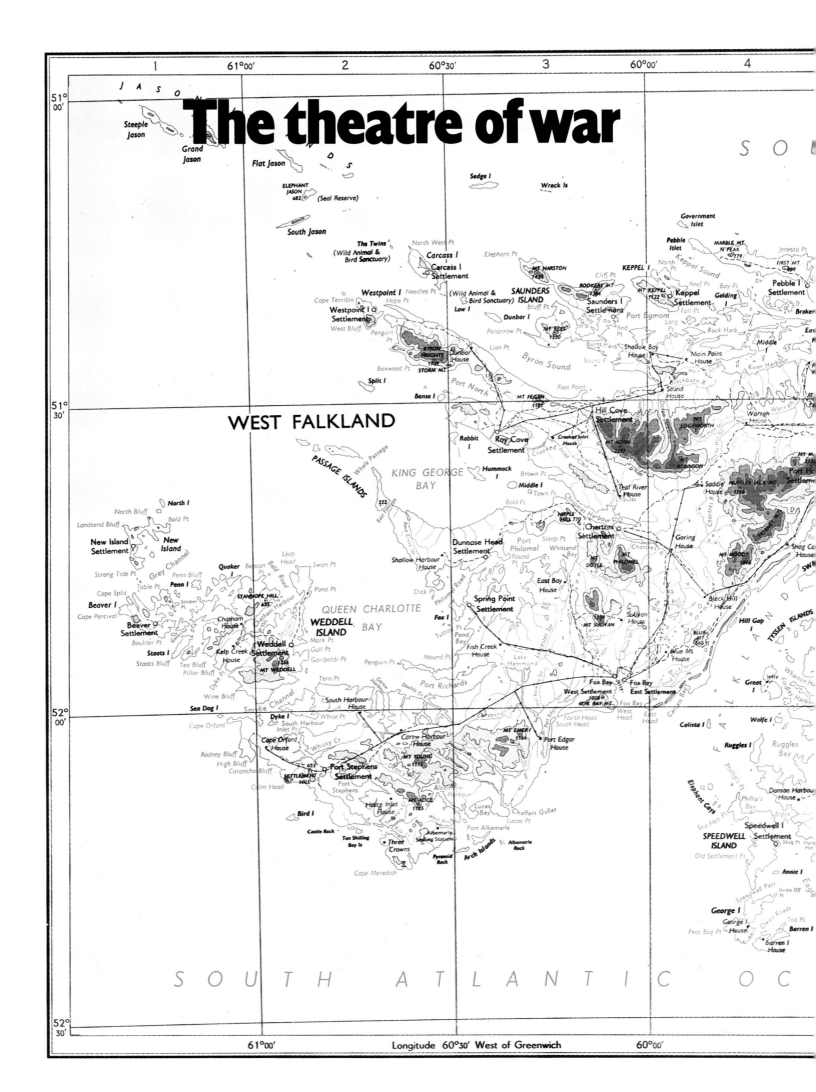

The theatre of war

WEST FALKLAND

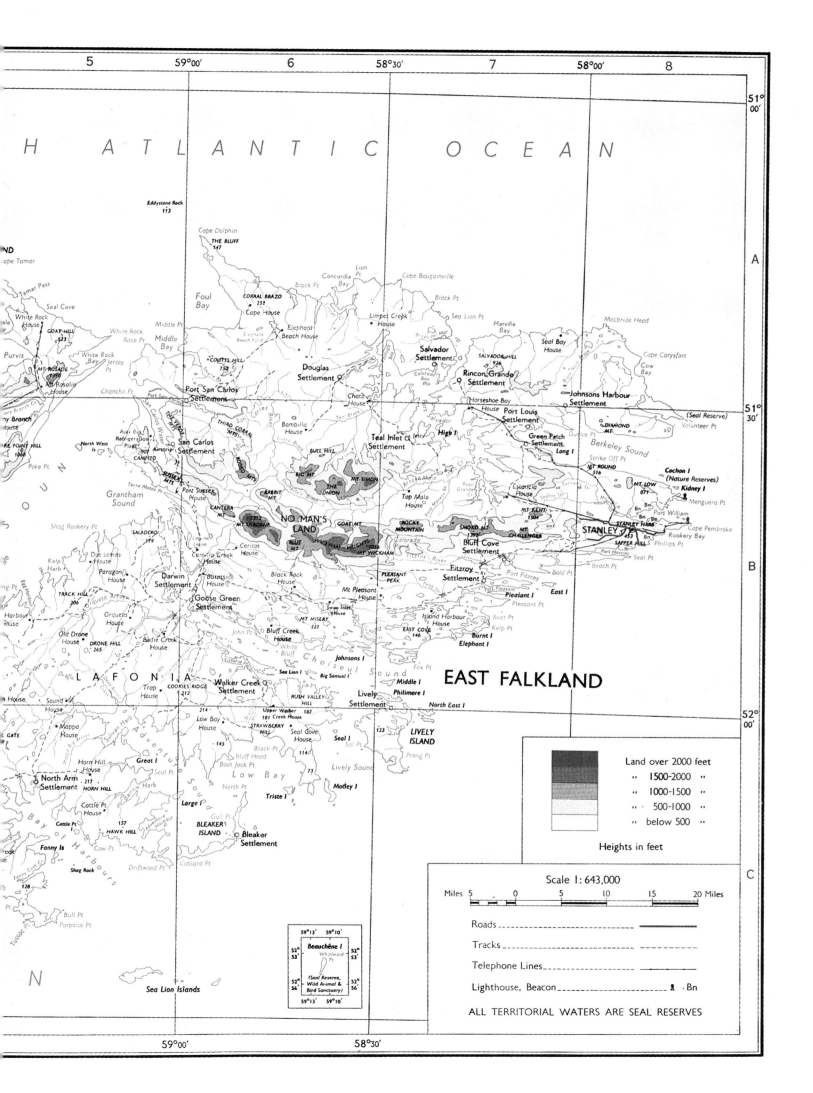

EAST FALKLAND

SOUTH ATLANTIC OCEAN

Scale 1 : 643,000

Miles 5   0   5   10   15   20 Miles

Land over 2000 feet
" 1500-2000 "
" 1000-1500 "
" 500-1000 "
" below 500 "

Heights in feet

Roads _____
Tracks _____
Telephone Lines _____
Lighthouse, Beacon _____ ☌ · Bn

ALL TERRITORIAL WATERS ARE SEAL RESERVES

Beauchêne I
Whirlwind Pt
(Seal Reserve, Wild Animal & Bird Sanctuary)

LAFONIA

Sea Lion Islands

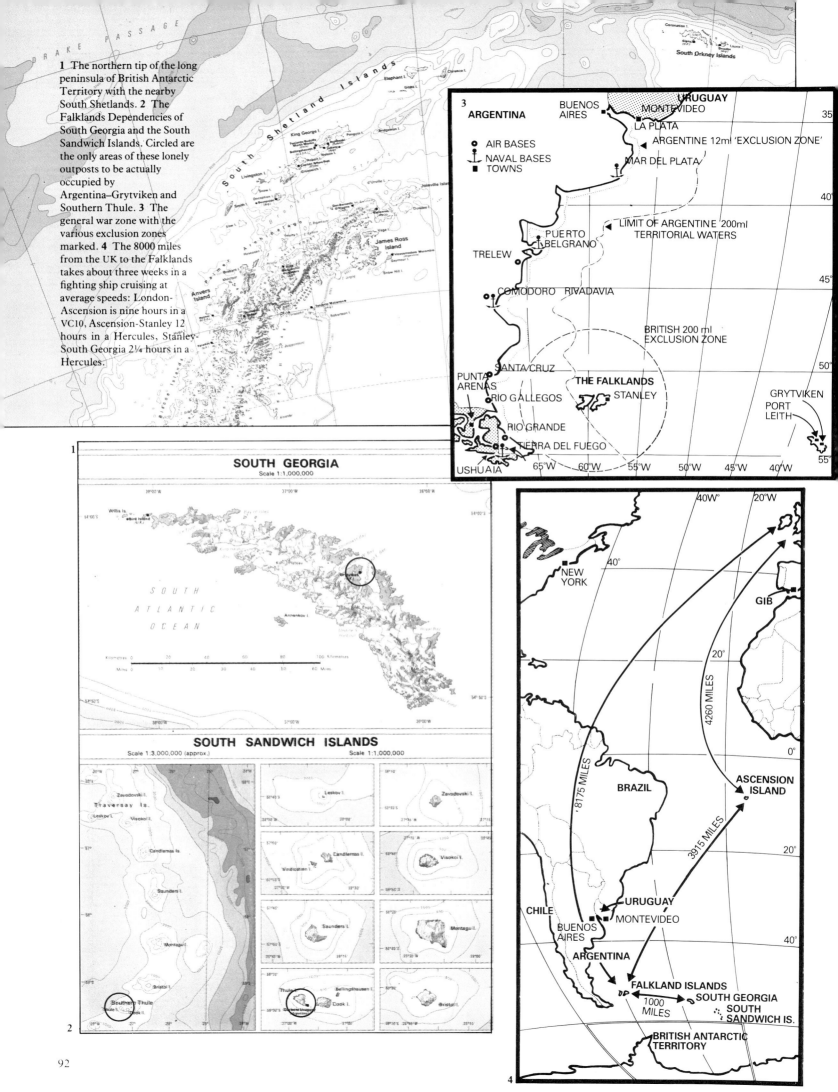

1 The northern tip of the long peninsula of British Antarctic Territory with the nearby South Shetlands. 2 The Falklands Dependencies of South Georgia and the South Sandwich Islands. Circled are the only areas of these lonely outposts to be actually occupied by Argentina–Grytviken and Southern Thule. 3 The general war zone with the various exclusion zones marked. 4 The 8000 miles from the UK to the Falklands takes about three weeks in a fighting ship cruising at average speeds: London-Ascension is nine hours in a VC10, Ascension-Stanley 12 hours in a Hercules, Stanley-South Georgia 2¼ hours in a Hercules.

SOUTH GEORGIA
Scale 1:1,000,000

SOUTH SANDWICH ISLANDS
Scale 1:3,000,000 (approx.)   Scale 1:1,000,000

The Falklands Stamps

A lasting footnote to the Falklands War is recorded in the philately of the islands. The stamp war has raged for years, the stamps themselves reflecting claim and counter-claim. And philately accounts for much of the Islands' income.

While the world's attention was fixed on the headline events of the Falklands War, the Post Office at Port Stanley, together with the world of stamp collecting at large, experienced some upheavals all their own: not the first, but certainly the most dramatic development in a philatelic war between Argentina and Great Britain that has been raging for decades.

Ever since 1936, when a map

The 'Malvinas treatment'

appeared showing the Falklands as part of Argentina, Argentine stamp issues have claimed both the islands and their dependencies. Regular 'British' issues, needless to say, were seen as philatelic usurpers and more than once, Encotel, the Argentine Post Office, refused to recognize their validity. In fact, between 1967 and 1972, the Buenos Aires Post Office would not accept letters posted in the islands.

On 5 April 1982, when Encotel staff took over the Port Stanley Post Office, the existing Falklands issues were locked away.

Top *Argentine 5000 pesos, 1982*
Above *Encotel cover issued day after San Carlos landings.*

Letters already in the mails, awaiting the weekly flight out, were given the 'Malvinas treatment'. Their Falklands stamps were crossed through and the envelopes marked with a special cancel with the legend '9409 Islas Malvinas'. Falkland postal clerks were undeterred: as a mark of passive resistance, they up-ended the date stamp and put the Argentine flag below, rather than above the date.

From 8 April, Argentine stamps were the official postage of the islands. Two weeks later, two million Argentine 1700 pesos

stamps were issued in Buenos Aires, defiantly overprinted 'Las Malvinas Son Argentinas'.

Stamp collectors specializing in the Falklands, even then a large group, were fascinated. This represented a superb new scenario for their albums, and Falklands issues, always popular for their interest and attractive design, became some of the most sought-after stamps of the day.

Argentine definitive with 'The Malvinas are Argentine' overprint, April 1982.

Unlike Argentina, which issues stamps in millions, the Falklands produce four to five high-quality sets a year providing, as stamps should, a miniature insight into island life and history. This preserves collector-interest and also contributes to Falklands'

finances, about 14 per cent of the island's gross domestic product deriving from the sale of its stamps.

Falkland commemoratives have featured sheep farming (1976), kelp and seaweed (1977) or the opening of Stanley Airport (1979). The British connection is often emphasized.

The larger definitive sets give fuller treatment of Falklands history, and in the case of the centenary definitive of 1933, have actually prompted it. The 1933 set was the first to show anything beyond the head of the reigning British monarch, and there were fierce Argentine objections to two of the stamps in particular: the 3d, showing East and West Falklands with all their English place names, and the £1, showing a head and shoulders portrait of George V. This set remains the most valuable set of all Falklands issues, fetching £2500 or more. The latest definitive set (11 stamps) was issued on 3 January 1983 to mark the 150th anniversary of Britain's occupation.

The Argentine map issue of 1936, issued in reply to its 1933 definitives, preluded Peron's propaganda campaign after 1947 to lay claim to the Falklands and dependencies.

After Peron was deposed in 1955, his Malvinas stamp policy went on unabated and over the next quarter-century, most such

isues featured the segment of Antarctica which Argentine claimed, quite illegally. Most were 'map' stamps, showing the islands as Argentine territory.

The 'Islas Malvinas son Argentinas' overprint of 1982 broke with this insinuating approach, and on 10 June, only four days before their surrender, the Argentines came out with their most blatant and largest

Falklands life reflected in its stamps.

stamp ever, a 5000 pesos map issue, 3½ in wide.

Falklands stamps as such have never deigned to refute any of the Argentine claims, but the war necessarily left its philatelic legacy: the giant £1 'rebuilding' stamps issued in July 1982.

On 14 June 1983, the first anniversary of the liberation, a four stamp Falklands set appeared honouring the Army, Royal

Emphasizing the British link—victory in World War 2 and the Churchill birth centenary, 1974.

Navy, Royal Air force and Merchant Navy. It was also the anniversary of the early morning departure of Encotel staff from Stanley. With them went their 9409 Islas Malvinas cancel—philatelically terminated.

Insignia and Units

SHIPS AND THEIR COMMANDING OFFICERS

HMS *Active*
Cdr PCB Canter, RN

TYPE 21 FRIGATE

HMS *Alacrity*
Cdr CJS Craig, DSC, RN

TYPE 21 FRIGATE

HMS *Ambuscade*
Cdr PJ Mosse, RN

HMS *Andromeda*
Capt JL Weatherall, RN

TYPE 21 FRIGATE

HMS *Antelope*
Cdr NJ Tobin, DSC, RN

DESTROYER

HMS *Antrim*
Capt BG Young, DSO, RN

Commander Task Force 317 and 324 Admiral Sir John Fieldhouse, GCB, GBE

Air Commander Air Marshall Sir John Curtiss, KCB, KBE, CBIM, RAF

Land Forces Deputy Major General Sir Jeremy Moore, KCB, OBE, MC and Bar. From 21 May: Lieutenant General Sir Richard Trant, KCB

Flag Officer Submarines Vice Admiral PGM Herbert, OBE

Commander Task Group 317.8 Rear Admiral Sir John Woodward, KCB

3 Commando Brigade Royal Marines Brigadier JHA Thompson, CB, OBE, ADC

5th Infantry Brigade Brigadier MJA Wilson, OBE, MC

Commodore Amphibious Warfare Commodore MC Clapp, CB

HMS *Ardent*
Cdr AWJ West, DSC, RN

HMS *Argonaut*
Capt CH Layman, DSO, MVO, RN

HMS *Arrow*
Cdr PJ Bootherstone, DSC, RN

HMS *Avenger*
Capt HM White, RN

HMS *Brilliant*
Capt JF Coward, DSO, RN

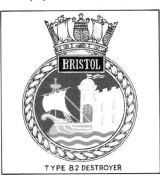

TYPE 82 DESTROYER

HMS *Bristol*
Capt A Grose, RN

HMS *Broadsword*
Capt WR Canning, DSO, ADC, RN

HMS *Cardiff*
Capt MGT Harris, RN

TYPE 42 DESTROYER

HMS *Coventry*
Capt D HART-DYKE, MVO, RN
HMS *Dumbarton Castle*
Lt-Cdr ND Wood, RN

ICE PATROL SHIP

HMS *Endurance*
Capt NJ Barker, CBE, RN

HMS *Exeter*
Capt HM Balfour, MVO, RN

ASSAULT SHIP

HMS *Fearless*
Capt EJS Larken, DSO, RN

DESTROYER

HMS *Glamorgan*
Capt ME Barrow, DSO, ADC, RN

TYPE 42 DESTROYER

HMS *Glasgow*
Capt AP Hoddinott, OBE, RN

HMS *Hecla*
Capt GL Hope, RN

HMS *Herald*
Capt RIC Halliday, RN

AIRCRAFT CARRIER

HMS *Hermes*
Capt LE Middleton, DSO, RN
– 800 Naval Air Squadron
Lt-Cdr AD Auld, DSC, RN
– 826 Naval Air Squadron
Lt-Cdr DJS Squier, AFC, RN

SURVEY SHIP

HMS *Hydra*
Cdr RJ Campbell, RN

ASSAULT SHIP

HMS *Intrepid*
Capt PGV Dingemans, DSO, RN

AIRCRAFT CARRIER

HMS Invincible
Capt JJ Black, DSO, MBE, RN
– 801 Naval Air Squadron
Lt-Cdr ND Ward, DSC, AFC, RN
– 820 Naval Air Squadron
Lt-Cdr RJS Wykes-Sneyd,
AFC, RN

HMS Leeds Castle
Lt-Cdr CFB Hamilton, RN

HMS Minerva
Cdr SHG Johnson, RN

HMS Penelope
Cdr PV Rickard, RN

HMS Plymouth
Capt D Pentreath, DSO, RN

DESTROYER TYPE 42

HMS Sheffield
Capt JFTG Salt, RN

HMS Yarmouth
Cdr A Morton, DSC, RN

HM SUBMARINES

FLEET SUBMARINE

HMS Conqueror
Cdr CL Wreford-Brown, DSO, RN

HMS Courageous
Cdr RTN Best, RN

HMS Onyx
Lt-Cdr AP Johnson, RN

HMS Spartan
Cdr JB Taylor, RN

HMS Splendid
Cdr RC Lane-Nott, RN

HMS Valiant
Cdr TM le Marchand, RN

ROYAL FLEET AUXILIARIES AND THEIR MASTERS

RFA Appleleaf
Capt GPA MacDougall, RFA

RFA Bayleaf
Capt AET Hunter, RFA

RFA Blue Rover
Capt JD Roddis, RFA

RFA Brambleleaf
Capt MSJ Farley, RFA

RFA Engadine
Capt DF Freeman, RFA

RFA Fort Austin
Cmdre SC Dunlop, CBE, DSO, RFA

RFA Fort Grange
Capt DGM Averill, CBE, RFA

RFA Resource
Capt BA Seymour, RFA

ROYAL FLEET AUXILIARY

RFA Olmeda
Capt AP Overbury, OBE, RFA

RFA Olna
Capt JA Bailey, RFA

ROYAL FLEET AUXILIARY

RFA Pearleaf
Capt J McCulloch, RFA

ROYAL FLEET AUXILIARY

RFA Plumleaf
Capt RWM Wallace, RFA

ROYAL FLEET AUXILIARY

RFA Regent
Capt J Logan, RFA

RFA Stromness
Capt JB Dickinson, OBE, RFA

RFA Tidepool
Capt JW Gaffrey, RFA

ROYAL FLEET AUXILIARY

RFA Tidespring
Capt S Redmond, OBE, RFA

Sir Bedivere
Capt PJ McCarthy, OBE, RFA

ROYAL FLEET AUXILIARY

Sir Galahad
Capt PJG Roberts, DSO, RFA

Sir Geraint
Capt DE Lawrence, DSC, RFA

ROYAL FLEET AUXILIARY

Sir Lancelot
Capt CA Purtcher-Wydenbruck,
OBE, RFA

ROYAL FLEET AUXILIARY

Sir Percivale
Capt AF Pitt, DSC, RFA

ROYAL FLEET AUXILIARY

Sir Tristram
Capt GR Green, DSC, RFA

ROYAL MARITIME AUXILIARY SERVICE SHIPS AND THEIR MASTERS

RMAS Goosander
Capt A MacGregor

RMAS Typhoon
Capt JN Morris

SHIPS TAKEN UP FROM TRADE, THEIR MASTERS AND SENIOR NAVAL OFFICERS

MV *Alvega*
Capt A Lazenby

MV *Anco Charger*
Capt B Hatton

MV *Astronomer*
Capt HS Braden
Lt-Cdr R Gainsford, RN

SS *Atlantic Causeway*
Capt MHC Twomey
Cdr RP Seymour, RN

SS *Atlantic Conveyor*
Capt I North, DSC
Capt MG Layard, CBE, RN

MV *Avelona Star*
Capt H Dyer

MV *Balder London*
Capt KJ Wallace

MV *Baltic Ferry*
Capt E Harrison
Lt-Cdr GB Webb, RN

MV *British Enterprise III*
Capt D Grant
Lt-Cdr BEM Reynell, RN

MV *British Avon*
Capt JWM Guy

MV *British Dart*
Capt JAN Taylor

MV *British Esk*
Capt G Barber

MV *British Tamar*
Capt WH Hare

MV *British Tay*
Capt PT Morris

MV *British Test*
Capt TA Oliphant

MV *British Trent*
Capt PR Walker

MV *British Wye*
Capt DM Rundell, OBE

SS *Canberra*
Capt W Scott-Masson, CBE,
Capt CPO Burne, CBE, RN

MV *Contender Bezant*
Capt A Mackinnon,
Lt-Cdr DHN Yates, RN

MV *Elk*
Capt JP Morton, CBE
Cdr AS Ritchie, OBE, RN

MV *Europic Ferry*
Capt CJC Clark, OBE
Cdr AB Gough, RN

MV *Fort Toronto*

Capt RI Kinnier

MV *G A Walker*
Capt EC Metham

MV *Geestport*
Capt GF Foster

GS *Iris*
Capt G Fulton
Lt-Cdr J Bithell, RN

MT *Irishman*
Capt W Allen

MV *Laertes*
Capt HT Reid

MV *Lycaon*
Capt HR Lawton
Lt-Cdr DJ Stiles, RN

MV *Norland*
Capt M Ellerby, CBE
Cdr CJ Esplin-Jones, OBE, RN

MV *Nordic Ferry*
Capt R Jenkins
Lt-Cdr M St JDA Thorburn, RN

RMS *Queen Elizabeth II*
Capt P Jackson
Capt NCH James, RN

TEV *Rangatira*
Capt P Liddell,
Cdr DH Lines, RN

MV *Saint Edmund*
Capt MJ Stockman
Lt-Cdr AM Scott, RN

RMS *Saint Helena*
Capt MLM Smith

MT *Salvageman*
Capt AJ Stockwell

MV *Saxonia*
Capt H Evans

MV *Scottish Eagle*
Capt A Terras

MV *Shell Eburna*
Capt JC Beaumont

MV *Stena Inspector*
Capt D Ede,
Capt PJ Stickland, RN

MV *Stena Seaspread*
Capt N Williams,
Capt P Badcock, CBE, RN

MV *Strathewe*
Capt STS Household
Lt-Cdr RH Hewland, RN

MV *Tor Caledonia*
Capt A Scott
Lt-Cdr JG Devine, RN

SS *Uganda*
Capt JG Clark
Cdr AB Gough, RN

Senior Medical Officer:
Surgeon Capt AJ Rintoul, RN

MV *Wimpey Seahorse*
Capt M Slack, OBE

MT *Yorkshireman*
Capt P Rimmer

MINESWEEPING TRAWLERS (*TAKEN UP FROM TRADE AND COMMISSIONED*)

HMS *Cordella*
Lt-Cdr MCG Holloway, RN

HMS *Farnella*
Lt RJ Bishop, RN

HMS *Junella*
Lt M Rowledge, RN

HMS *Northella*
Lt JPS Greenop, RN

HMS *Pict*
Lt-Cdr DG Garwood, RN

COMMANDER BRITISH FORCES SUPPORT UNIT, ASCENSION ISLAND, UNTIL 17 JUNE 1982

Capt R McQUEEN, CBE, RN

ROYAL MARINE UNITS AND COMMANDING OFFICERS
3 Commando Brigade
Headquarters and Signal Squadron
Royal Marines
Maj RC DIXON, RM

40 Commando Royal Marines
Lt Col MPJ HUNT, RM

42 Commando Royal Marines
Lt Col NF VAUX, DSO, RM

45 Commando Royal Marines
Lt Col AF WHITEHEAD,
DSO, RM

Lt Col NF Vaux, DSO

Lt Col AF Whitehead, DSO

Commando Logistic Regiment
Royal Marines
Lt Col IJ HELLBERG, OBE, RCT

3 Commando Brigade Air
Squadron Royal Marines
Maj CP CAMERON, MC, RM

1st Raiding Squadron Royal
Marines
Capt FIJ BAXTER, RM

Special Boat Squadron
Royal Marines
Maj JJ THOMSON, OBE, RM

3 Commando Brigade Air Defence
Troop Royal Marines
Lt IL DUNN, RM

Y Troop Royal Marines
Capt GD CORBETT, R Signals

Field Records Office Drafting
and Records Office Royal Marines
Capt JR HANCOCK, RM

The Band of Her Majesty's Royal
Marines Commando Forces
Capt JM WARE, LRAM, RM

Lt Col IJ Hellberg, OBE

The Band of Her Majesty's Royal
Marines Flag Officer
3rd Flotilla
WO 2(B) T ATTWOOD, LRAM,
ARCM, RM

ARMY UNITS AND COMMANDERS
Two troops The Blues and Royals
Capt RAK FIELD, RHG/D

4th Field Regiment Royal Artillery
(Less one battery)
Lt Col GA HOLT, RA

12th Air Defence Regiment Royal
Artillery (Less one battery)
Lt Col MC BOWDEN, RA

29th Commando Regiment Royal
Artillery
Lt Col MJ HOLROYD-SMITH,
OBE, RA

Elements 43 Air Defence Battery,
32nd Guided Weapons Regiment
Royal Artillery
Capt RC DICKEY, RA

Elements 49th Field Regiment
Royal Artillery
Maj RT GWYN, RA

Elements Royal School of Artillery
Support Regiment
Maj MH FALLON, RA

Lt Col MIE Scott, DSO

Elements 33 Engineer Regiment
Capt B LLOYD, RE

36 Engineer Regiment
(Less one squadron)
Lt Col GW FIELD, MBE, RE

Elements of 38 Engineer Regiment
Maj RB HAWKEN, RE

59 Independent Commando
Squadron Royal Engineers
Maj R MACDONALD, RE

Elements Military Works Force
Lt Col LJ KENNEDY, MBE, RE

Elements 2 Postal and Courier
Regiment Royal Engineers
Maj I WINFIELD, RE

Elements 14th Signal Regiment
Capt GD CORBET, R Signals

Elements 30th Signal Regiment
Maj WK BUTLER, R Signals

5th Infantry Brigade Headquarters
and Signals Squadron
Maj ML FORGE, R Signals

Elements 602 Signal Troop
Warrant Officer II (Yeoman
of Signals)
JF CALVERT, R Signals

2nd Battalion Scots Guards
Lt Col MIE SCOTT, DSO, SG

1st Battalion Welsh Guards
Lt Col JF RICKETT, OBE, WG

Maj CPB Keeble, DSO

1st Battalion 7th Duke of
Edinburgh's Own Gurkha Rifles
Lt Col DP de C MORGAN, OBE, 7GR

2nd Battalion The Parachute
Regiment. Commanded in turn by
a) Lt Col H JONES, VC, OBE, Para
b) Maj CPB KEEBLE, DSO, Para
c) Lt Col DR CHAUNDLER, Para

3rd Battalion The Parachute
Regiment
Lt Col HWR PIKE, DSO, MBE, Para

Elements 22nd Special Air Service
Regiment
Lt Col HM ROSE, OBE,
Coldm Gds

656 Squadron Army Air Corps
Maj CS SIBUN, AAC

Elements 17 Port Regiment Royal
Corps of Transport
Lt JGD LOWE, RCT

Elements 29 Transport and
Movements Regiment Royal Corps
of Transport
Lt DR BYRNE, RCT

Elements 47 Air Despatch
Squadron Royal Corps of Transport
Maj RC GARDNER, RCT

407 Troop Royal Corps of
Transport
Lt JP ASH, RCT

Lt Col HWR Pike, DSO, MBE

Elements of The Joint Helicopter
Support Unit
Corp J ELLIOT, RCT

16 Field Ambulance Royal Army
Medical Corps
Lt Col JDA ROBERTS, RAMC

Elements 19 Field Ambulance
Royal Army Medical Corps
Cap JT GRAHAM, RAMC

Elements 9 Ordnance Battalion
Royal Army Ordnance Corps
Maj RBP SMITH, RAOC

81 Ordnance Company Royal
Army Ordnance Corps
Maj GMA THOMAS, RAOC

10 Field Workshop Royal Electrical
and Mechanical Engineers
Maj AD BALL, REME

Elements 70 Aircraft Workshops
Royal Electrical and Mechanical
Engineers
Staff Sergeant MJ EMERY, REME

Elements 160 Provost Company
Royal Military Police
Cap AK BARLEY, RMP

6 Field Cash Office Royal Army
Pay Corps
Maj RF CLARK, RAPC

601 Tactical Air Control Party
(Forward Air Controller)
Maj MM HOWES, RRW

602 Tactical Air Control Party
(Forward Air Controller)
Maj AS HUGHES, RWF

603 Tactical Air Control Party
(Forward Air Controller)
Flight Lt G HAWKINS, RAF

ROYAL AIR FORCE
Senior Royal Air Force Officer,
Ascension Island, and Commander
British Forces Support Unit
Ascension Island from 17 June
Group Capt JSB PRICE, CBE,
ADC, RAF

FLYING SQUADRONS AND COMMANDERS
1 (F) Squadron Harrier GR3
Wing Cmdr PT SQUIRE, DFC,
AFC, RAF

Detachments of
10 Squadron VC10
Wing Cmdr OG BUNN, MBE, RAF

18 Squadron Chinook HC1
Squadron Leader
RU LANGWORTHY, DFC,
AFC, RAF

24 Squadron Hercules C1
30 Squadron Hercules C1
47 Squadron Hercules C1
70 Squadron Hercules C1
Squadron Ldr MJ KEMPSTER,
RAF (4-17 Apr 82).
Squadron Ldr JRD MORLEY, RAF
(18 Apr-11 May 82).
Squadron Ldr NCL HUDSON,
BA, RAF
(12 May-23 Jul 82).

29 Squadron Phantom FGR2
Squadron Ldr RWD TROTTER,
RAF

42 Squadron Nimrod Mk.1
Wing Cmdr DL BAUGH, OBE, RAF

44 Squadron Vulcan B2
50 Squadron Vulcan B2
101 Squadron Vulcan B2
Squadron Ldr AC MONTGOMERY,
RAF

Group Capt JSB Price, CBE, ADC

55 Squadron Victor K2
57 Squadron Victor K2
Wing Cmdr DW MAURICE-JONES,
RAF (18-21 Apr 82)
Wing Cmdr AW BOWMAN, MBE,
RAF (22 Apr 82)

120 Squadron Nimrod Mk.2
201 Squadron Nimrod Mk.2
206 Squadron Nimrod Mk.2
Wing Cmdr D EMMERSON,
AFC, RAF

202 Squadron SAR Sea King
Flight Lt MJ CARYLE, RAF

*Squadron Ldr RU Langworthy
DFC, AFC*

ROYAL AIR FORCE REGIMENT
3 (Regiment) Wing Headquarters
Unit and 15 (Regiment) Squadron
Detachment Field Squadron
Wing Cmdr TT WALLIS, RAF

63 (Regiment) Squadron (Rapier)
Squadron Ldr
IPG LOUGHBOROUGH, RAF

Support Units
Tactical Communications Wing
Tactical Supply Wing
No 1 EOD Unit

Badges

ROYAL NAVY

ROYAL MARINES

ROYAL FLEET AUXILIARY

ROYAL CORPS OF SIGNALS

SCOTS GUARDS

WELSH GUARDS

ROYAL ARMY CHAPLAIN'S DEPARTMENT

ROYAL CORPS OF TRANSPORT

ROYAL ARMY MEDICAL CORPS

ROYAL ARMY EDUCATIONAL CORPS

ROYAL PIONEER CORPS

INTELLIGENCE CORPS

ARMY CATERING CORPS

Every man and woman who served during the Falklands War is represented by one of these badges, or was attached to a service whose badge is illustrated below. The great number of units involved is proof, if any were needed, that Operation Corporate was a team effort.

THE BLUES AND ROYALS

ROYAL REGIMENT OF ARTILLERY

CORPS OF ROYAL ENGINEERS

THE PARACHUTE REGIMENT

7th DUKE OF EDINBURGH'S OWN GURKHA RIFLES

SPECIAL AIR SERVICE REGIMENT

ARMY AIR CORPS

ROYAL ARMY ORDNANCE CORPS

ROYAL ELECTRICAL AND MECHANICAL ENGINEERS

ROYAL MILITARY POLICE

ROYAL ARMY PAY CORPS

QUEEN ALEXANDRA'S ROYAL ARMY NURSING CORPS

WOMEN'S ROYAL ARMY CORPS

ROYAL AIR FORCE

SOUVENIR PAGE 15

Daily Mail

TUESDAY, JUNE 15, 1982 — 17p

Chrissie
HER OWN STORY
—see Pages 24, 25

- Large numbers of Argentine soldiers threw down their weapons. They are flying white flags over Port Stanley

—Mrs Thatcher in the Commons

VICTORY!

IT IS all but over. The Argentines ran up white flags over Port Stanley la... began talks...

Maggie's sweetest moment

By GORDON GREIG, Political Editor

MRS THATCHER will never

Daily Mail 17p

WEDNESDAY, JUNE 16, 1982

VICTORY IN THE FALKLANDS
12-page pull-out

...the general as Port Stanley cele...

THE DAY OF LIBERATION

land forces on the islands walked down the main street—pitch black because of a generator failure—to a torchlit barn-like store where about 125 islanders welcomed him with a huge cheer. I'm Jeremy Moore, he said. 'I'm sorry it took us three weeks to get here.'

JUBILANT Falkland Islanders — and one hardly old enough to understand—greet their liberator, Major-Gen. Jeremy Moore. Minutes after accepting the Argentine surrender, the commander of Britain's

Wom... hand. Th... shoulders... Fellow. E... and toas... They... was wha...

Femail 10, Diary 13, TV Guide 28, 29, Letters, Strips & Stars 30, City, Casino 32, Chris Evert Lloyd Story 34, F...

DAILY EXPRESS

Thursday June 17 1982 — 17p — Weather: Cloudy

V=F

FEAR FOR PRISONERS AS ARG...

Starving sick and ABANDO...

...iana's...

DAILY MIRROR

Friday, June 18, 1982 — 14p

Knifeman held at the Palace

By JACK McEACHRAN

A MAN armed with a knife was arrested at Buckingham Palace last night.
The intruder, who got into the Palace grounds by clambering over a fence, was spotted by a uniformed guardsman.

The soldier challenged the man and detained him until other guardsmen and police arrived. The knife was found when he was searched. None of the royal family was staying at the palace at the time. A man will appear in court today.

Curfew as captured troops go on rampage

FALKLAND islanders were under curfew in Port Stanley last night following a riot by Argentine prisoners.

And British troops were confined to barracks after a two local time in attempt to ease the tension. Falklanders were told to stay indoors after 8pm. Paratroopers set up checkpoints and patrolled the capital and pubs were shut down indefinitely.

The Argentines erupted in an orgy of looting and drink after being taunted by islanders.

Military police had to threaten to open fire before they cooled down.

An angry fireman who helped clean up the mess said later: "They're bastards, worse than pigs. They've turned Port Stanley into a rubbish tip."

The prisoners ran out of control as they were being marched in batches of 500 from the airfield to the docks.

There they were to have been ferried to the cruise ship Canberra to be repatriated.

Blaze

But on the way scuffles broke out with islanders who came out of dockside bars to taunt them.

The march broke up after a rumour started that the ship was leaving without them.

Some started running towards the docks. Others ran into another street and released two parked brakes on an empty armoured car, sending it careering wildly downhill.

Another group broke into the Globe Store, a building rented by their British army for their stockpiling equipment

From ALASTAIR McQUEEN
In Port Stanley

and food — fooled it, and set it alight.

As a volunteer force of armed men fought to control the blaze, another fire started in the squash club next to the junior club where the Argentine forces had been billeted.

Food

I heard one shot fired and then a Military Police corporal warned the prisoners through one of their English-speaking officers: "If you don't calm down we'll open fire."

Sodden kit bags and blankets were dumped in the streets in their thousands, and sacks of food floated in the puddles.

At the junior school the classrooms were littered with rotting food, remains of meals, boxes of spilled ammunition and hand grenades.

While I looked around, ammunition flames reached the window.

Next door, in the senior school, Royal Marines were clearing up using gallons of disinfectant to wash the floors and walls and to kill the sickening stench.

Other troops organised working parties of prisoners and made them clear up

GALTIERI GETS THE BOOT

ARGENTINA'S military supremo, General Leopoldo Galtieri, was booted from power last night.

And moderate air force boss Lami Doso emerged as favourite to take over as the new president.

For the moment, the job is being done by a stopgap president, General Alfredo Saint Jean.

Galtieri's role as military commander-in-chief has been taken over by General Cristino Nicolaides, a hard-liner who is opposed to free political parties.

From PAUL CONNEW in Buenos Aires

The appointment of Lami Doso—a national hero after the battle for peace over the Falklands—would be a big boost for hopes that ending the military conflict with Britain and launching a diplomatic campaign.

This appears to be the majority view among Argentina's military leaders—which saw Galtieri, 55, was ousted. He shaked his political career on continuing the fighting.

Galtieri's disgrace came with his exchange with the prisoners and the 10,000 Argentine p...

He rejected a plea f... so that the prisoners... mainland by a fleet o...

Many of the pris... with malnutrition,... diarrhoea, trench-fe...

Galtieri said that... soldiers sent home... want them back w... unloading them at...

There were hi... them back through...

—Pages 4 and 5

VICTORS and VANQUISHED

DAILY ...

Tuesday...

The w... flags... flying... in Sta...

WHITE flags were flyin... Thatcher told the C... they will soon be repl...

The Prime Ministe... reception as she ... announce an immediate...

Talks are now in ... render of Argentine f... Our troops have been ... in self-defence.

As the cheering ... Michael Foot and De... rose to congratulate t... armed forces.

THE STANDARD
Wednesday, June 16, 1982. 15p

**Galtieri warned: Hu[n]
could die of disease,**

'YOU
MEN
FAC
DISAST[E]

ADMIRAL WOODWARD—"If a choice has
to be made, defence comes first."

DAILY Mirror
Tuesday, June 15, 1982   14p

**FALKLANDS
ARE TAKEN**

VICTORY
•A[rge]ntine troop[s]

By ELLIS PLAICE in London and PAUL CONNEW in
Buenos Aires
BRITISH troops won the battle of the
[Falklands] last night. A ceasefire was
[...] [Arge]ntine

[...]SS
[...]CE OF BRITAIN
[I]GNORES PLEAS

Enemy troops eat in the open

[...]NED

[E]XPRESS
THE VOICE OF BRITAIN
17p   • Weather: Sunny spells

Ceasefire
as enemy
surrender
their arms

BRITISH troops stormed triumphantly into the
outskirts of Port Stanley last night with the routed
enemy under pressure to surrender.

An Argentinian military spokesman said in
Buenos Aires that the two countries had agreed a
cease-fire in the Falklands until 2 p.m. London
time today.

Argentine resistance collapsed after 48
hours of intensive military action by the 6,000
British. Enemy troops laid down their arms.

Surrender talks started between the Argentine garrison commander, General Mario
Menendez, and Brigadier John Waters.

Earlier the junta authorised him to negotiate
with the British "as long as the honour of the
armed forces is not compromised."

General Moore.

[...]rrender — full story: Pages 2, 3, 5, 6, 7 and Centre Pages

DAILY STAR
TUESDAY, JUNE 15th, 1982   14p   Printed in London

**WHITE FLAGS
FLY OVER
PORT STANLEY**

**THEY
DID NOT
DIE IN
VAIN**

Heroes
of the war

THEY unselfishly gave their
lives, like so many others.
Left, from top : Captain Ian
North, of Atlantic Conveyor;
Acting Leading Marine
Stephen White, of HMS
Ardent; Staff Sergeant Jim
Prescott, Bomb Disposal;
Private Stephen Illingsworth,
Parachute Regiment. Right,
from top : Lieutenant Colonel
H. Jones, Parachute Regi-
ment; Bosun John Dobson,
of Atlantic Conveyor; Able
Seaman Sean Hayward, of
HMS Ardent; Lance Corporal
Anthony Cork, Parachute
Regiment. Bottom centre :
Lt.-Cmdr. Gordon Batt.

Galtieri [...] quit after jan[...]

**LEADER
NATION IN[...]**

FAL[K...]
VI[CTORY]

BRITAIN [...]
brate w[...]
dropped [...]
and fl[...]
white [...]
Port [...]
surren[der...]
midnight [...]

More than 220 [...]
Servicemen died in the
Falklands campaign. The
nation knows it was not
in vain.

Special reports:
Pages 2, 3, 4, 5, 6 and 7

£50,000 BIRT[H...]

Picture credits
Aerospatiale: 24 (t), 26/7
Alvis Ltd. (Coventry): 50 (t, b)
Aviation Photographs Int: 14 (t1, tr), 15, 34/5 (t), 64/5
Avions Marcel Dessault: 34 (t), 34/5 (b)
G. L. Bound: 102/3
British Aerospace: 8/9, 21, 42 (t), 43 (t, b), 64 (t)
Camera Press: 51 (t, b)
C.O.I.: 10/11 (b), 22 (b)
Commando Forces News Team: 96 (c, b)
Richard Cooke: 10/11 (t), 20 (t)
Dept of Defence/USA: 64 (b)
Lt. M. Duck RM: 6, 36 (br), 48/9 (c), 66
Euromissile: 36 (b), 37 (tl, tr), 39 (tr)
Express Newspapers: 22 (t), 23 (t), 97 (cl)
Graeme Harris: 98/9
John Hillelson Agency: 20 (b), 25 Sygma
P.O. Peter Holdgate: 1, 48/9 (t, b)
Lockheed: 14/15
Marconi: 31
Mars/Crown copyright/MOD: 4, 54, 60/1
MOD: 55 Navy, 97 (cr, br) Studio D, Blackpool
Dept. of National Savings: 82 (c), 83 (t), 84, 85 (b), 86, 87 (b)
Press Association: 7 (Martin Clever), 34 (b), 82 (l, r), 83 (b), 85 (tl, tr), 87 (t), 96 (t), 97 (t, bl)
Rex Features: 65 (b)
The Royal Mint: 84, 87 (c), 89
Short Bros. Ltd: 42 (b)
David Smith: 94/5
Soldier Magazine: 39 (tl)
Spink & Son Ltd.: 88
Sygma/Stuart Franklin: 17
Vickers Shipbuilding Ltd: 16 (t, b)
Westland Helicopters: 30, 30/1

Artwork
British Aerospace: 42 (t), 43 (t, b)
Crown copyright/DOS: 90/1 (1966), 92
K. Lyles: 67—80
Sarson & Bryan: 22/3, 24/5 Aerospaciale, 32/3, 38/9, 40/1
British Aerospace, 56/7, 58/9, 62/3
Martin Streetly: 12/13, 23 (b), 24 (b), 32 (t), 38 (t), 63 (t)

Index

Page numbers in *italic* refer to the illustrations and captions.